THE PURSUIT OF
DESTINY

THE PURSUIT OF
DESTINY

A History of Prediction

PAUL HALPERN

PERSEUS PUBLISHING
Cambridge, Massachusetts

Many of the designations used by manufacturers and sellers to distinguish their products are claimed as trademarks. Where those designations appear in this book and Perseus Books was aware of a trademark claim, the designations have been printed in initial capital letters.

A CIP record for this book is available from the Library of Congress.
ISBN: 0-7382-0095-6
Copyright © 2000 by Paul Halpern

Perseus Books is a member of the Perseus Books Group.
Find us on the World Wide Web at http://www.perseuspublishing.com

Text design by Jeff Williams
Set in 11-point AGaramond

First printing, September 2000
1 2 3 4 5 6 7 8 9 10—03 02 01 00 99

For Aden,
wise beyond his years

CONTENTS

ACKNOWLEDGMENTS

I wish to acknowledge the invaluable assistance of my colleagues at the University of the Sciences in Philadelphia, including Philip Gerbino, Barbara Byrne, Charles Gibley, Elizabeth Bressi-Stoppe, Bernard Brunner, Stanley Zietz, Nancy Cunningham, Harriet Gomon, John Martino, Jude Kuchinsky, Paul Angiolillo, Salar Alsardary, Greg Manco, David Kerrick, Durai Sabapathi, Barbara Reilly, Roy Robson, Elizabeth Adams, Robert Boughner, Michael Dockray and David Traxel. I appreciate Peter Hoffer's advice on translating a German passage.

Thanks to Martin Gardner for many useful references and recommendations about prediction paradoxes, futurology, and related matters, to J. Doyne Farmer for many insightful comments about the science of forecasting, to Paul E. Rapp for suggestions about complexity in neuroscience and psychiatry, to James Randi for a fascinating discussion about pseudoscience, to Saul Perlmutter for interesting information about his supernova research, to John Holland for valuable information about his complex systems research, and to Jeff Robbins for his encouragement. Thanks to Amanda Cook for her insightful editorial suggestions and general advice about this book, and to the staff of Perseus Books for their assistance.

I would also like to thank my family and friends for their support, including Stanley, Bernice, Richard, Alan, Kenneth, Anita, and Esther Halpern, Joseph and Arlene Finston, Michael Erlich, Fran Sugarman, Scott Veggeberg, Marcie Glicksman, Fred Schuepfer, Pam Quick, Simone Zelitch, and Dubravko Klabucar. Above all, I'm grateful for the endless love and guidance of my wife Felicia and sons Eli and Aden, who have enriched my life in countless ways and helped me to pursue my destiny.

Yesterday This Day's Madness did prepare
Tomorrow's Silence, Triumph, or Despair
—THE RUBA'IYAT OF OMAR KHAYYAM
(translated by Edward FitzGerald)

INTRODUCTION

The Shape of Things to Come

All things are full of weariness;
a man cannot utter it;
the eye is not satisfied with seeing,
nor the ear filled with hearing.
What has been is what will be,
and what has been done
is what will be done;
and there is nothing new under the sun.
Is there a thing of which it is said,
"See this is new"?
It has been here already,
in the ages before us.
There is no remembrance of former things,
nor will there be any remembrance
of later things yet to happen
among those who come after.

—ECCLESIASTES 1.8–11

Yearning for the Future

From the moment of conception until the instant of annihilation, the currents of time pull us slowly and steadily forward. As much as we strive to break free, to witness what lies ahead, we are completely encapsulated in the ark of the present moment. Seemingly powerless to alter the pace of our journeys through time, we simply ride them out, never knowing with certainty what will transpire next.

Nonetheless, the desire to foretell, understand, and ultimately explore the future is an integral part of what makes us human. We are the only species on Earth that demonstrates vivid interest in the times to come. Our visions of the days, months, and years ahead guide our actions in a unique fashion. As acclaimed science writer Martin Gardner writes:

> Knowing the future clearly helps one make decisions about the present. If I knew the stock market would crash later this year, I would sell all my stocks. If I knew a certain airplane would crash, I wouldn't take it. If I knew a woman would poison me if I married her, I wouldn't propose. From the beginning of history people have naturally wanted to know the future, and of course there are ways to make reasonable guesses.[1]

The forward-looking behavior of humans, based on complex mental representations of the future, contrasts sharply with the instinctual responses of other living creatures. Most animals react solely on the basis of their own aversions and desires—avoiding negative stimuli and adopting behavior that will best satisfy their basic needs such as hunger and thirst. Fleeing from loud noises and electric shocks and drawing nearer to the pungent smell of food, they respond impulsively to the environment around them. Often such fast reactions are correct; if a squirrel stops to think about the screech of an approaching car, it might get run over. But sometimes, on the basis of these quick responses, animals driven merely by instinct fall too easily into traps. Fish, for instance, jump at the bait, and then they're literally hooked.

Humans too have instincts. These are appropriate in a wide range of situations, from fleeing imminent dangers to responding correctly to social cues. However, uniquely we can override such behavior when perceived long-term benefit outweighs immediate gains or losses. Rather than spending all their money on lavish meals and expensive vacations, for instance, a couple might decide to save up for a house. Fighting his natural desire to sleep, a student might elect to stay up all night studying extra hard for an exam.

Although higher animals sometimes act on the basis of short-term consequences, they cannot meaningfully compare long-range alternatives. Rather than planning ahead, they are instinctively responding to remembered experiences. Based on what has happened to them before, they avoid actions leading to dangers (or take actions leading to rewards) shortly down the road. For example, a dog might learn that if it topples over garbage cans when its master is not around, it will be punished once the master returns.

Only humans, however, can act imaginatively on the basis of long-term planning—scoping out potential problems and possibilities months, years, even decades ahead. Such facility requires the skill of visualization: the formulation and manipulation of detailed mental models (internal representations) of likely alternatives. The capacity to create, compare, and update these images of possible futures requires well-honed thinking processes that often involve the use of language. To consult with others—and even to talk to oneself—one needs to employ the symbolic shorthand that words and phrases provide. Other animals lack centers in the frontal lobes of their brains for higher-level cognition, such as for language in the fullest sense of the word (manipulating symbols to convey novel ideas). Consequently these less developed species lack the capability of meaningfully considering future outcomes.

One might conjecture, then, that the human capacity for basing behavior on long-range forecasts has come about through biological selection. The greater the ability of a species to predict what will happen, the more successful it is at avoiding catastrophes. Like the propensity of plants to turn toward light, our quest for the future forms an integral part of our means to survival.

Progress in Prediction

The human ability to exploit knowledge of the future has been far from static. Our large brains provide us with the necessary "hardware" to perform the task of visualization. The "software" required to perform this task with increasing accuracy and efficiency, however, has developed considerably over the millennia. As the full potential of science has unfolded, the methods for assessing prospects and pitfalls have correspondingly advanced.

In ancient times, envisioning the future was far from being a precise, scientific task. Soothsayers and priests scanned the skies, examined the entrails of sacrificial animals, and interpreted their own dreams in attempts to forecast the state of things to come. Great rulers, from Cyrus to Caesar, relied on these and other methods of divination for their military, governmental, and personal decisions. Exalting their counselors in times of triumph, they admonished, shunned, or sometimes even executed them in times of unforeseen hardship.

Today's predictive experts—laboratory scientists, think-tank researchers, government advisers, and others—have more reason to worry about being sacked than sacrificed. Though their methods are radically different from those of antiquity, they share with the ancients the anxiety that their prognostications will wholly miss the mark. Although accurate forecasting can lead to auspicious business decisions, sound political policies, and creative measures to prolong lives, errors in prediction can result in economic catastrophe, social turmoil, and needless death.

The willingness to pay, and pay dearly, for superior forecasts has driven mathematicians, scientists, and engineers to fashion increasingly sophisticated predictive technology. From furnishing advanced knowledge about the weather to farmers to offering detailed information about possible enemy strategies to generals, whenever there has been a market for predictive information, experts have aspired to match all demands with ready supplies of data. To best their competitors, forecasters have sought superior algorithms and more efficient computational devices, advancing the science of the future ever forward.

Though the science of forecasting has advanced considerably since the days of augury, its limitations have grown correspondingly more obvious. Some of these limits are technological in nature, reflecting the limitations of computational systems. Making a scientific prediction requires detailed information about the past and present behavior of a system, a model for how this behavior might be extrapolated into the future and the computational means for performing this extrapolation. Naturally, the faster the computer's central processing unit, the more precise the results and the more efficiently they'll be achieved.

Other limits to prediction are more fundamental in nature, and are correspondingly more troubling. Sometimes better computation is not enough to overcome the built-in constraints posed by natural law. Specifically, radical changes in scientific thinking over the past hundred years—in fields such as quantum theory, chaos theory, and complexity theory—have demonstrated the critical roles that chance and uncertainty play in the dynamics of physical processes, and have consequently revealed essential limitations to what can be known.

The Journey Ahead

The story of destiny's impact on humanity and humanity's influence on destiny has been cast and recast in various forms through the centuries, from Shakespearean dramas to modern doctoral dissertations. The twin questions of how much can we know about the future and how much can we change about the future have been the impassioned focus of countless debates among thinkers from all nationalities, backgrounds, and disciplines.

This book is written from a physicist's point of view, and therefore emphasizes the prevailing scientific notions at the turn of the twenty-first century. Books have been written concentrating on economic, political, and social forecasting; this is not one of them. My main intention is to examine how relativity, quantum theory, chaos theory, complexity theory, cosmology, and other aspects of modern physics have affected our long-standing quest to comprehend our fate.

Before we begin our journey to the lofty heights of contemporary theoretical discussion, we need to prepare ourselves well—nourishing ourselves with a hearty basic background and equipping ourselves with essential concepts and terminology. To that aim, we shall start with an overview of the history of prediction, in both its scientific and nonscientific forms. Focusing on cosmology, the wellspring of predictive science, we'll examine how humankind's desire to map the rhythms of the celestial spheres led to remarkable discoveries about the laws of nature. These findings, in turn, resulted in the development of classical mechanics—which many thinkers once believed could theoretically predict anything.

Along the way, we'll encounter noted natural philosophers who have made vital contributions to the study of prediction: Pythagoras, Plato, Aristotle, Cicero, Augustine, Kepler, Galileo, Newton, and many others. Throughout our voyage, these luminaries shall serve as our guides, helping us to interpret and place in context modern debates about the nature of destiny.

We'll also meet figures from the "darker side" of the history of prediction—those who have specialized in supernatural prophecy. From the Pythian priestesses who dispensed the wisdom of Apollo through cryptic utterances to the Kabbalists, who pondered the secret meaning of the Scriptures through mathematical codes, to Nostradamus, who contemplated the tapestry of tomorrow through deep meditation, and, finally, to current mystical believers who have claimed prescience through a variety of methods, we'll examine age-old attempts to glean extraordinary insight into the shape of the future.

Remarkably, we'll find close historical connections between those who sought scientific understanding and those who embraced supernatural beliefs. In many cases—Pythagoras, Kepler, and Newton, for example—scholarly achievements were spurred on, at least in part, by the passion for mystical knowledge about the cosmos. In other examples—Nostradamus comes to mind—trained scientists in their later years have claimed extraordinary spiritual powers, or—in the case of Michael Drosnin in his unraveling of the so-called "Bible Code"–have claimed computational insights into the omniscient gaze of God. Apparently a taste for one kind of prediction often leads to an appetite for other types of premonitions.

We will then look at how contemporary physical theory envisions space, time, and the future—from quantum gravitational models of temporal progression to self-organizing systems operating on the edge of chaos. We will examine paradoxes and puzzles concerning bizarre implications of modern science, exploring hypothetical notions such as wormholes and time travel. If the world, as some have pondered, is a multidimensional labyrinth, full of shortcuts, bifurcations, and dead ends, then the concept of destiny is manifestly far from straightforward.

Seeking solid and familiar ground, we will then tread the solemn roads of social, technological, and medical forecasting, concentrating on how modern scientific notions have enhanced these fields. We'll look, for example, at how theories of chaos and complexity have helped enable researchers to model the onset of heart attacks and strokes, how neural network models and other computer algorithms have offered businesses critical tools for planning out their inventories, and how constructs based on Darwinian selection have helped explain why some outdated technologies survive, while others are tossed aside. As we'll discuss, commercial prognostication has become a booming industry—centered in its own Silicon Valley: Santa Fe, New Mexico—where scientists trained in physics, biology, and other fields have refocused their talents on developing and running predictive software for businesses.

In exploring the accomplishments of computational science, we'll also discuss one of its valuable lessons: as researchers Kurt Gödel, Alan Turing, Gregory Chaitlin, and others have shown, all mathematical schemes possess built-in limits. In other words, no self-consistent system for understanding the world and its future can be complete. That's why certain problems in forecasting would run forever on the most powerful computers and never find resolution.

Even in today's technological age, human insight still provides the critical ingredient for attempts to anticipate the future. We'll examine techniques researchers have developed to pool the collective wisdom of experts in evaluating the likelihood of various far-reaching scenarios. These methods, however, clearly have their limitations. Looking back on forecasters' images of the year 2000, we'll show how scattered "hits" are unfortunately far outnumbered by "misses." Although some predictions made decades ago have proven correct, particularly concerning medical and genetic technology, many others have landed far from their marks. For example, whereas projections about moving sidewalks on every street, robot servants in every household, and picturephones on every desktop never materialized, nobody anticipated the ubiquity of e-mail. To understand why so many long-range social and technological forecasts prove false, we'll investigate the limits posed by the sheer unpredictability of free-thinking minds.

Finally, in our journey's fitting culmination, we will examine the ultimate application of predictive science: astronomers' attempts to unlock the mysteries of the final stages of the universe. Remarkably, researchers have developed theories of what might happen trillions of years from now. With far-reaching models such as these, traditional questions in forecasting—such as what are the chances of rain tomorrow—seem mundane in comparison. Yet it is the mark of human adaptability that our thoughts might range in an instant from simple curiosity about the next day's weather to deep reflection about the supreme destiny and purpose of the cosmos.

The drive for future knowledge is powerful indeed. Today, we can turn to science in attempts to satisfy our thirst for information about the times ahead. The ancients, however, had no such luxury. Without access to scientific methods, to answer their questions about what may or may not happen, they came to depend on supernatural means of prognostication. In ancient Greece, for example, all important decisions were rendered in consultation with an oracle, or spiritual mouthpiece. And no source of godly wisdom and divine foreknowledge was more exalted in the classical world than the oracle of Delphi.

1

ANCIENT AUGURIES

And, truly, what of good
ever have prophets brought to men?
Craft of many words,
only through evil your message speaks.
Seers bring aye terror, so to keep men afraid.

—AESCHYLUS (AGAMEMNON)
translated by Edith Hamilton

The Oracle of Delphi

In a dark, subterranean alcove of a majestic temple in Apollo's holy city, an olive-skinned woman with long, graying hair gazes up at the ceiling. Though her youth has clearly faded, she is dressed in the colorful costume of a maiden. Sitting on a three-legged stool, she is surrounded by a small group of robed men, gazing intently upon her like eager children.

The pungent aroma of burned laurel leaves fills the air, rendering each observer a bit light-headed. No one speaks, at first, for fear of disturbing the solemnity of the ceremony. Finally, when the time is judged to be most auspicious, one of the men recites a prayer to Apollo, and entreats the god of wisdom to answer a simple, but vital, query.

The woman's face suddenly turns pale. Her lips begin to contort, as if possessed. With blank expression she mutters a series of strange, barely intelligible phrases with otherworldly tones. In monotonous ca-

1

dence, her bizarre admonitions strain out one by one—like wet sand dripping slowly through a sieve.

If she were on a crowded Athenian street or marketplace, surely she would have been mistaken for mad. Like the legendary misunderstood prophetess Cassandra, her utterances would have gone untranslated, her garbled warnings unheeded, until it was too late for action. But here, in the Temple of Apollo in the city of Delphi, her every sound and movement is duly recorded. Priests transform her frenzied phraseology into metered verse. Properly recast, in form chosen for ready interpretation, empires might rise or fall on these very words.

It is supremely ironic that in ancient Greece, the birthplace of science and the cradle of logic, the mysterious power of prophecy held sway for hundreds of years. In the time known as the Archaic period, from the eighth to the sixth centuries B.C., the preponderance of wars that were fought, agreements that were made, and political decisions that were rendered were initiated and guided in consultation with supernatural forces. The ears of kings, priests, warriors, and traders alike were attuned to those said to speak for the gods.

Unquestionably, the divine conduit held highest in esteem was the oracle of Delphi. In ancient Greek legend, Delphi was the center of the universe. Zeus, the king of the gods, once saw fit to release two eagles, one from the East and the other from the West. When the two met in Delphi, he placed a stone there, marking it the world's navel. An extraordinary place with an exalted history, Delphi was the holiest of holy cities, the focus of godly truth and wisdom.

Greek mythology speaks of colossal battles fought over this powerful site. Originally, Delphi was known as "Pytho." It belonged to Gaea, goddess of the earth, and was guarded by her child, Python, a serpent. One day, Apollo, a young and virile god who possessed supreme knowledge of the future, saw fit to slay Python. Drawing his blade, he lay claim to Gaea's cherished sanctuary.

After committing this murder, Apollo left the city and disguised himself as a dolphin. He leaped aboard a Cretan ship, captured its crew, and forced them to take him back to Delphi. Together they conquered the city, naming it after the Greek word for dolphin. Henceforth, Delphi became the main shrine of Apollo and the spiritual center of Greece.

FIGURE 1.1 *The theater of Delphi, part of the ruins of one of the greatest cities of the ancient world (courtesy of the Library of Congress).*

In the Greek mindset, a drink from the Delphic font of foreknowledge was seen as the elixir for great political and military power. Great leaders from throughout the classical world sent emissaries to Apollo's sacred city. Testing the Delphic waters with sundry queries ranging from the trivial to the monumental, they hoped to quaff a taste of cosmic truth. Fortified with insight into the future, the leaders felt well prepared for whatever projects they wished to undertake—peaceful affairs or warlike incursions.

In Greek tradition, at least once each generation a woman known as the Pythia was chosen to serve as the conduit of Apollo's powers. Though in the early years of this tradition youth were selected to fill this post, later, to reduce the possibility of corruption, only middle-aged women were picked. Typically, the Pythia was a simple, respectable, country-bred woman who elected to leave her family and take up a monastic existence.

One feature that marked each of the Pythian priestesses was her "enthusiasm." "Enthusiasm," from the Greek enthousiasmos, in its original sense means "possessed by a god." Her private hopes and yearnings considered unimportant—even antithetical to her role—fundamentally the Pythia was deemed merely a vessel through which Apollo might pour his precious future knowledge.

The sense in which Apollo dispelled his wisdom was widely understood to have sexual connotations. The Greeks believed that Apollo disseminated his foreknowledge only to women that he possessed. Once a woman became a Pythia, she offered herself freely to Apollo as a "lover," and agreed to eschew all other intimate relationships, even with her husband. If she abused the privilege of being Apollo's conduit—by denying him his due and offering herself to others—then she would almost certainly earn his wrath.

For those familiar with the tale of Cassandra, its dire message served, no doubt, as a warning to those who contemplated disobedience to their god. According to legend, Cassandra, the daughter of King Priam of Troy, took advantage of Apollo's loving generosity, forsook him by not reciprocating, and consequently suffered his wrath. In the epic of Agamemnon, which tells the story of the Trojan War, the playwright Aeschylus relates her sorrowful chronicle.

As the play describes, Apollo fell in love with the beautiful young Cassandra. To woo her, he bestowed upon her the supreme gift of magnificent prophetic powers. When she betrayed him by not becoming his lover, he became enraged, afflicting her with the curse that no one would ever heed her warnings. Thus, though she would know the horrors of the future, she would be powerless to change them.

Shortly thereafter, Cassandra was struck with a terrible premonition. Greek armies, led by the great soldier Agamemnon, were planning to invade her beloved city of Troy. She screamed out her warning cries, but was merely ridiculed. Soon it was too late; the invaders stealthily entered and occupied her kingdom.

When victory was complete, Agamemnon took Cassandra as his prize. He told her that he was going to take her back to his homeland. She warned him that if they went to Greece they both would be murdered. Again, her admonitions were ignored and they set sail.

Once they landed in Greece, Clytemnestra, Agamemnon's wife, soon found out about her husband's young and attractive new mistress. In a fit of jealousy, the enraged queen had them both slain. Apollo's revenge was complete; Cassandra had been forced to live her final years in anticipation of unavoidable doom.

In contrast to Cassandra, the women who prophesied from the Delphic temple of Apollo were well respected. Their warnings, if understood, were taken extremely seriously—there was little chance that they'd be ignored. Often they received precious gifts from individuals eager for their advice.

For example, when Croesus, the fabulously wealthy king of Lydia (from whom the expression "rich as Croesus" arose), was deciding whether or not to go to war with Persia, he paved his way for a speedy and meaningful answer from the oracle with magnificent presents. She gladly received the favors, and promptly advised him that if he embarked upon his campaign an empire would fall. He launched his war, but unfortunately, it turned out that she meant his empire! Lydia fell, not Persia.

Plutarch, the great Greco-Roman biographer, reports that the typical ceremony at Delphi took place one day each month (except during the three winter months, when the center, located high on Mount Parnassus, was often covered with snow). During this sacred "day of Apollo," pilgrims bringing vital queries were admitted to the temple. Slogans designed to sharpen the thoughts and purify the heart of each inquirer would greet him as he entered. "Know yourself" and "be temperate," inscribed upon the temple walls, were intended to salute him with Apollo's essential message of humility.

Before the inquirer was allowed to present his question, the Pythia and priests would prepare themselves. The Pythia would bathe in the sacred spring of Castalia, drink from its holy waters, chew laurel leaves (Apollo's sacred tree was the laurel), and consume a ritual cake. She would then descend into her divine chamber and sit on a three-legged stool, known as a tripod. The priests would also immerse themselves in Castalian waters and sacrifice a goat on the temple's altar. Finally, the pilgrim would bathe as well, and be brought into the Pythia's chamber.

Plutarch reports that the seated Pythia would breathe in vapors from cracks in the earth beneath her seat. She would soon act dazed and withdrawn as if she were in a trance. The worshiper would offer his query, the priests would present it to her, and she would respond in a strange, deep voice seeming to emanate from her belly. The priests would translate her answer into hexameter (Greek verse) form, and re-lay it back to the pilgrim.

Many scholars of the Roman and early Christian eras, attempting to debunk the oracle's mystique, suggested that the Pythia's frenzied responses were directly induced either by her inhaling the smoke or chewing the laurel leaves. They pictured her as a sort of madwoman in a perpetual stupor, engaged by the priests as their hapless proxy. The priests, these scholars implied, would usually disregard the Pythia's ramblings and fashion their versed answers in a manner they felt would best suit their own needs.

Some historians, ancient and modern, have even suggested that bribery played a role in the oracle's decisions. In their view, the free acceptance of gifts by priests during the ceremony could have subtly or even overtly affected their translation of the Pythia's words into verse. The scholar Demosthenes surmised that during the reign of Philip of Macedonia, the Pythia herself was paid to be under his control—"phillipized," he put it. Nevertheless, many other thinkers have extolled the wisdom and impartiality of the Pythia's decisions. If she or the priests were bribable, then why didn't Croesus, of unsurpassed wealth, get his way? Most likely, as in any profession, different priestesses possessed varying degrees of honesty, sensibility, and insight.

If the Pythia was chosen for her plain, simple background, why then did she act so strangely while she was delivering her predictions? In modern times, historians have attempted to settle the question of whether or not the Pythia's behavior was drug induced. Laurel leaves have been tested for their hallucinogenic properties; apparently none are discernable. The floor of the ruins of the temple has been searched for cracks and fissures. None have been found, leading scholars to conclude that perhaps there were no fumes.

In the absence of evidence of a pharmaceutical agent, many scholars have suggested that the Pythia's rantings were hypnotically induced instead. According to this popular view, while she was caught up in the ceremony, psychological factors, rather than chemical agents, brought her into her mesmerized state. Thus her strange speech patterns more closely resembled the less-than-coherent tones of someone talking right before they fell asleep, rather than the utter gibberish of someone thoroughly stoned.

In 1997, Wesleyan University geologist Jelle de Boer delivered a talk at a meeting of the Geological Society in London that seemed to resurrect the intoxication theory. He presented strong evidence that the ground beneath the base of the Temple of Apollo could well have been the site of seismic activity in ancient times.

Roadwork in the region of the temple ruins had exposed the substratum and made it possible for de Boer to investigate nearby areas previously unexplored. He found active geological faults to the east and west of the site, and smaller cracks to the north and south. Quite possibly, he concluded, unseen fissures existed under the temple as well, opening and closing whenever the earth shifted.

At the London conference, de Boer pointed out a mechanism by which the Pythia may have been exposed to noxious gases. During the height of Greek civilization, earthquakes may have cracked open the temple floor, producing the openings that Plutarch described. Through these fissures, chemical fumes, such as ethylene, methane, and hydrogen sulfide vapors—generated by the decay of hydrocarbons in the limestone strata beneath the site—may have seeped in. The reason contemporary scientists have not detected these cracks, de Boer speculated, is that more recent seismic activity has sealed them.

De Boer's collaborator, John Hale of the University of Louisville, believes that the gases helped the Pythia to enter her contemplative state. "The exhalations had the effect of putting the woman into the right spirit to make these prophecies, whether or not they actually produced intoxication,"[1] he says.

Whether she was conscious or hypnotized, fully cognizant or out of her mind, completely impartial or highly political, a proud spokeswoman for high ideals or a puppet manipulated by the priests, the Pythia's effect upon the classical Greek psyche cannot be overestimated. She was the siphon through which Greek civilization drank in the power of Apollo, invigorating and fortifying itself again and again. For the years in which Greece was at its height, the Pythia personified its image as the chosen nation of the gods and embodied its claim to be the hub of the cosmos. In essence, the wise madwoman (or mad wisewoman) who sat on the tripod was Delphi, and the cryptic, labyrinthine city that sat upon Mount Parnassus was Greece.

The Wellspring of Science

In an age in which the rambling utterances of oracles constituted the hallmark of official prognostication, a philosophical tradition emerged in Greece that ultimately proved far more significant for predictive science. The schools of Pythagoras, Plato, Aristotle, and others pioneered a bold new approach to understanding the future—the search for the basic patterns and movements of nature. Focusing their eyes on the flow of the heavens and their ears on the rhythms of music, the classical Greek philosophers pioneered the notion of anticipating the pulse of the future by examining the cadence of the present. Their introduction of the notion of describing the world through numbers and their emphasis on geometric symmetry have provided inestimable contributions to the science of mapping out the possibilities of tomorrow. For without deciphering the hidden codes of the present, by discovering the world's significant mathematical features, how might man expect to unravel the riddles of coming times?

Born on the island of Samos around 570 B.C., Pythagoras apparently was named in honor of the Pythia, or even after Apollo himself. True to his Apollonian legacy, he, like the Pythia, became a conduit of wisdom and a deliverer of mysterious knowledge. In Crotona, a city in a region of southern Italy colonized by the Greeks, he established a school, or "brotherhood," to teach and promote mystical, as well as mathematical, ideas. For him, the secrets of the occult and the patterns of numbers each provided essential keys for unlocking reality's sacred vault of prescient information. As a religious leader as well as a teacher, he recruited hundreds of disciples who agreed to obey him completely without question, submit to an ascetic vegetarian lifestyle, and study mathematics, music, and astronomy.

The merger of faith and reason was not unusual for that era. In ancient Greece, at the time of Pythagoras, lines between what we now call mysticism and what was to be known as science were far more blurred than they are today. Scholars of the motions of heavenly bodies or the properties of geometric figures were interested in these subjects mainly for what they revealed about man's purpose and destiny. To them, nature was a vast puzzle in which Apollo encrypted his secret portrait of the future, and those artful enough to rearrange the pieces might glimpse a precious vision of truth.

The key to Pythagorean thought is the concept of kosmos, a word likely coined by Pythagoras or one of his disciples to refer to the eternal order that underlies all things. Limitless and perfect in all respects, it possesses an essential harmony that appears in its purest form in astronomy, music, and mathematics. The astronomical vision of kosmos manifests itself in the Pythagorean notion that the stars, planets, Sun, and Moon revolve around a central fire (the dwelling place of divine forces) on concentric, transparent celestial domes. Their ceaseless, rhythmic motion—the so-called "harmony of the spheres"—brings on the succession of day and night, distinguishes the seasons, and sets the clockwork pace of earthly ritual.

Closely related to this celestial portrait of order is the Pythagorean concept of harmonia: musical harmony. Pythagoras has traditionally been attributed with the invention of the musical scale. In this system, ratios of the first four integers characterize relationships between sounds. Tones one octave apart are related to each other by the ratio 1:2. A 3:2 ratio denotes a fifth, and a 4:3 ratio, a fourth. Pythagoras and his followers found that music that followed this scheme sounded the most pleasing to the ear. Their sense of the aesthetic led them to conclude that harmonia reflected the underlying organization of the universe.

In the view of the Pythagoreans, what unites the musical and celestial realms with the prosaic world of people and things is the mystery of numbers. A much-quoted phrase, "number is all," expresses well their philosophy of nature. Not just that mathematics describes the cosmos, as contemporary science believes; rather, in their system, mathematics is the cosmos. Like an intricate tower laid out stone by stone and brick by brick, they held that the cosmic edifice is constructed of various combinations of integers. The essential role that modern chemistry assigns to the universe's atomic constituents—hydrogen, helium, lithium, and so on—the Pythagoreans attributed to the counting numbers: one, two, three, and so forth.

Though the Pythagorean approach to the cosmos assigns numerical values to natural properties and thereby provides a nascent example of mathematical analysis, it is far removed from forecasting in the modern sense. Unlike modern scientific prediction methods, it fails to distinguish numbers from the things they represent. To sense quantitative change, quantities must be allowed to change; if they are inextricably connected to what they designate, they cannot. Therefore, Pythagorean theory cannot analyze the dynamics of an evolving system.

The Pythagorean school eventually met an unfortunate demise. The residents of Crotona, disturbed by the group's secrecy, drove its members away and burned down its buildings. Pythagoras fled to Metapontum, another part of southern Italy, where he spent the waning years of his life. According to one story, he was murdered there. After the death of its founder, the remnants of the sacred brotherhood dispersed through the Greek world. Only thanks to the favorable writings of subsequent philosophers such as Plato did the golden legacy of Pythagoras survive.

Plato, born in Athens around 428 B.C., was indisputably one of the greatest philosophers who ever lived. His institute for research and learning, the Academy, founded in 387 B.C., served as a prototype for intellectual advancement and training from which modern universities draw their roots. His students, particularly Aristotle, mapped out extensively, in a hitherto unknown manner, the fundamental attributes of nature. Though more often than not these intellectual endeavors led to incorrect conclusions, they constituted a comprehensive source of ideas with which later, truer notions might be compared.

Not only did Plato's Academy serve as the birthplace for modern science in general, it provided, in particular, an incubator from which modern scientific forecasting eventually came forth. For forecasting theory to develop, the mathematical aspects of Pythagorean theory needed to be severed from its mystical parentage. Plato did not cut this umbilical cord himself; conditions were not yet right to do so. Years later, after forecasting eventually acquired the capacity to survive as an independent entity, his followers were able to perform this task.

One of the most important contributions of Plato was his exploration of the properties of time. He was the first to distinguish everlasting, cyclical, and ephemeral aspects of temporal passage, and to explain how an immortal designer might set the clock of mortal existence in motion. In *Timaeus*, he describes how the Creator, an unchanging, eternal entity, manufactured cyclical time as an imperfect copy of perpetual time:

> When the father and creator saw the creature which he had made moving and living, the created image of the eternal gods, he re-

joiced, and in his joy determined to make the copy still more like the original; and as this was eternal, he sought to make the universe eternal, so far as might be. Now the nature of the ideal being was everlasting, but to bestow this attribute in its fullness upon a creature was impossible. Wherefore he resolved to have a moving image of eternity, and when he set in order the heaven, he made this image eternal, but moving accord to number, while eternity itself rests in unity; and this image we call time.[2]

In other words, because the Creator could not make everything in his created universe perpetual like himself, he did the next best thing; he took his own image and copied it again and again. More precisely, he swung it round and round, in the form of rotating celestial orbs. For Plato saw the clockwork rhythms of the planets as the perfect expression of the singularity of eternity reflected in the multiple mirrors of cosmic repetition.

Despite the Pythagorean roots of his model, Plato's way of looking at the world represents several more steps down the road to modern prediction. Though, like Pythagoras, he often refers to abstract qualities, Plato also makes ample reference to physical characteristics such as time and space. He theoretically admits the possibility of change, even complex change, but then argues (wrongly) that circularity comprises the natural pattern of the universe.

Another of Plato's innovations is his theory of Forms. Instead of arguing that "number is all," and that the world of substance is directly produced by the actions of numbers, he proposes that material objects constitute the incomplete versions of perfect entities called Forms. The observed world is a shadow play; the real action occurs behind the curtain, in the ceaseless drama of the forms. Once again, Plato's ideas are abstract, but at least they decouple natural objects from numbers, and numbers from mysticism.

For centuries, historians have generally looked askance at Plato's theory of Forms, deeming it, at best, a curious aspect of classical philosophy, and, at worst, a dangerous delusion. Arthur Koestler, in his classic history of early cosmology, *The Sleepwalkers*, singles it out as a terrible misstep that set back science for generations.

"When reality becomes unbearable, the mind must withdraw from it and create a world of artificial perfection," he writes. "Plato's world of pure Ideas and Forms, which alone is to be considered as real, whereas the world of nature which we perceive is merely its cheap Woolworth copy, is a flight into delusion."[3]

However, as it has turned out, the notion of representing observed physical properties with hidden mathematical entities has become an essential feature of quantum mechanics. Plato's theory splendidly anticipated an important aspect of modern physics.

Koestler also criticizes Plato's failure to imagine noncircular patterns and his "fear of change," which he sees as misguided tendencies subsequently exacerbated by Aristotle. True, Plato as a man of his times was attracted to the notion that all things run in cycles. Virtually all philosophers of the ancient world embraced this idea, seen as a clear extension of the periodicity of the natural world (expressed, for instance, in the rhythm of the seasons). Following the inclinations of his age, as well as his own predisposition toward perfect systems, Plato fervently believed in reincarnation of all things; to him there truly was nothing new under the sun.

Plato was wholly correct in assuming that the planets move in cycles. He erred, however, in believing that the planets move in circles. Lacking access to accurate astronomical data, it is hardly surprising that he did not realize—as Kepler ultimately demonstrated—that planetary orbits are elliptical. Considering the undeveloped state of observational science that characterized all periods before the modern era, Plato's circular theory was quite understandable.

After Plato died in 347 B.C., his movement became deeply divided. Sharply different interpretations of his works generated impassioned debate among his disciples whether to proceed further in the direction of tangibility or to veer back toward pure abstraction. Some followers, such as Aristotle, argued for movement toward an empirically based natural philosophy. Others, such as Plato's nephew Speusippis, pushed for a virtual return to Pythagorean numerical mysticism. The successors of these disparate tendencies constituted powerful historical forces long after Plato was gone.

Aristotle, born in the Macedonian town of Stagira 384 B.C., became a student of Plato's Academy when he was seventeen. Founding his own school, the Lyceum, in 335 B.C., he wrote dozens of books on an extraordinarily broad range of topics, from astronomy to ethics and from political behavior to zoology (though some might lump the latter two fields together). Unfortunately, only a fraction of these writings have survived—enough to demonstrate the impressive versatility of one of the greatest natural philosophers.

One of Aristotle's primary missions was to cast off the shroud of superstition cloaking the face of science and unmask the underlying principles of nature. With meticulous observation and flawless logic he aspired to provide a subject that had long been veiled in the rhetoric of numerological mysticism, with a far more natural and perceptible outward appearance. The visage of natural philosophy, he felt, should be plain, lucid, and self-evident, a clear representation of the elegant simplicity and rationality of nature itself.

In his *Metaphysics*, Aristotle criticized Pythagorean cosmology, with its abstract references to mathematics and music. Instead, he imagined the cosmos as a physical system: a concentric series of orbiting objects, each revolving around Earth. In contrast to Plato, Aristotle believed that the celestial bodies constituted reality itself; they were not mere shadows of Forms. Set into motion by a "prime mover," the planets, Sun, Moon, and stars would each follow their natural tendency (as Aristotle thought) to move in circles forever.

Contrast the Aristotelian picture of astral revolutions around Earth with the Pythagorean image of celestial motion around a central hearth of divine power. By placing Earth at the center of the universe, nestled in the throne where once sat the gods, Aristotle, in one bold step, disconnected the physical from the spiritual realms. No longer would scientific prediction need to derive its legitimacy by saddling itself to religious aspirations; from that point on, it could seek out its own route. The path from Aristotle to Newton, and then onward to modern forecasting, was clear and direct; it is unfortunate indeed that human science languished in the desert for centuries until reason prevailed.

Though each derived from interpretations of their mentor's theories, the sharp, rational, empirically grounded perspective of Aristotle contrasted sharply with the obscure, numerological visions of most of his contemporaries. Under the direction of Speusippis, Plato's doctrine of Forms became reconstituted as a doctrine of numerical relationships. To Aristotle's chagrin, the new leadership of the Academy appeared more interested in the exposition of abstractions than in the understanding of nature. Like the self-absorbed philosophers satirized by

Swift, they embraced introspection, not observation, as their primary wellspring of knowledge. Their zeal to use their minds' eyes, not their own eyes, in discerning the patterns of the divine, lingered until medieval times and infused successor movements such as Neoplatonism, Gnosticism, and Kabbalism. Hence the tendency, throughout the Middle Ages, for groups to try to predict the future through numerological deduction, rather than through scientific measurement.

Aristotle's analytical approach, embraced from time to time by new philosophical admirers, gradually assumed the status of Dickens's Mrs. Havisham: dusty, dejected, and virtually dead. Jilted by its courtiers and later trapped in the attic of Church teachings, its bridal veils yellowed, its complexion faded, and its teeth rotted for centuries. Only in the Renaissance were these naturalistic methods fully revived. And then, only in the Enlightenment, fully 2,000 years after their debut, were Aristotle's theories, in their entire original charm and splendor, finally surpassed.

Modern prediction owes a great debt to classical philosophy. Without the contributions of Pythagoras, Plato, Aristotle, and others, humankind's quest to understand its world and its destiny likely would have proceeded along a radically different path. The Greek philosophers' aspiration to fathom the cosmos through numbers, shapes, and rhythms represented an endeavor of extraordinary scope and importance. Unquestionably, in the history of forecasting, their efforts comprised a titanic leap beyond auguries, lots, interpretation of dreams, and other traditional means of divination.

Dreams, Schemes, and Madness

Plato referred to supernatural prophecy as a type of madness that takes over the body during times of illness. He proposed that the center for divination was located in the liver and that this function was inactive in times of health. Only when the liver was full of impurities might someone acquire the ability and desire to mutter predictions. He wrote:

> The authors of our being . . . gave to the liver the power of divination, which is never active when men are awake or in health; but when they are under the influence of some disorder or enthusiasm then they receive intimations, which have to be inter-

preted by others who are called prophets, but should rather be called interpreters of prophecy.

Even when the oracle of Delphi and other recognized oracles were at their height of influence, in classical Greece and throughout the ancient world, prophetic "enthusiasm" was hardly confined to official centers of worship. From the time that Greek philosophy was at its peak to the age of Rome, and even in the Christian era, the fervent desire for information gleaned from supernatural sources spurred innumerable practitioners to engage in untold varieties of divination.

As long as scientific methods of prediction are imperfect (and they probably will always have shortcomings, because of built-in limits that we'll discuss), soothsayers of different sorts will profit by attempts to satisfy people's hunger for knowledge of the future. Today's limits to forecasting certainly provide ample frustration; who has not wished they could know the weather weeks in advance to plan an outing or a wedding? Imagine the frustration, then, in ancient times, when scientific forecasting was almost wholly undeveloped, and largely confined to discerning simple celestial patterns. Virtually any malady had unknown prognosis; any long-range problem, an unknown solution. The gap, in that point in history, between the desire for information and science's capacity to fulfill that want was inestimably more pronounced. Hence the soothsayer's potential audience was desperate, enormous, and willing to pay the price.

Aside from the oracles serving the classical Greeks, the most respected of traditional divinatory practices—then and now—is the pseudoscience of astrology. Astrology draws its esteem from its ties to the genuine science of the stars, purportedly representing the influence of the celestial bodies on worldly affairs. The Sun, the Moon, the planets, and constellations are supposed to exert direct influences on the human psyche, particularly at birth. In ancient times, this force was considered to be divine; today's astrologers will often argue that it is gravitational. Although many scientists have pointed out that when children are born the doctor that delivers them exerts a greater gravitational attraction than do the stars in space, the notion of astral influence remains a popular myth.

Because ancient peoples could predict very little about nature, those who could perform astronomical forecasts were especially revered and their observations were viewed with awe. To anticipate the behavior of the planets, to record eclipses, to plan calendars, and to render other kinds of celestial judgments were all considered holy tasks. It was only natural that individuals esteemed for their excellent astronomical prognoses were also called upon for political and personal advice. Thus astronomer, astrologer, and adviser constituted one and the same profession.

Astrologers served as counselors to the thrones of the Egyptian pharaohs, Chinese emperors, and Babylonian kings, among others. These royal advisers provided detailed reports and interpretations of the appearance of new stars, eclipses, comets, meteor showers, and other cosmic phenomena. They drew up detailed horoscopes and pointed out possible omens.

According to historical accounts, the Babylonians were particularly adept astrologers. Their records of solar and lunar eclipses, dating as far back as 700 B.C., are often remarkably accurate, sometimes timed to the nearest four minutes. Because of the precision of these annals, modern scientists have found them enormously valuable—a boon to understanding those times.

Richard Stephenson of the University of Durham, England, and Leslie Morrison, formerly of the Royal Greenwich Observatory in Cambridge, England, have collected more than 300 eclipse reports, from Babylonian, Chinese, Arab, and European records, and have used them to estimate changes in Earth's spin over time. These researchers have shown that in 500 B.C. the day was about fifty milliseconds shorter than it is today. Currently they are seeking to decipher Egyptian astrological records, carved in hieroglyphics, to press back their chronology even further.[4] Hence, early astrologers are still offering forecasts, though undoubtedly not in the way they imagined.

Aside from astrology, few of the ancient forms of divination are familiar to us today. With the rise of modern technology, contact with the natural environment has diminished, and interest in techniques that rely on observations of nature has consequently declined. Many early naturalistic methods, such as examination of animal entrails (haruspicy) and livers (hepatoscopy), interpretation of lightning, thunder, bird cries, and flight patterns (ornithomancy), gazing at smoke and fire (pyromancy), and noting the appearances of sacred animals (bears, eagles, serpents, etc.), faded into disuse ages ago. The American custom of observing woodchucks' shadows during Groundhog Day for signs of winter's end, home-grown meteorological predictions in farmers' almanacs, and other folk myths about the weather serve as present-day relics of such naturalistic traditions.

Other techniques that rely less on natural observation and more on chance and probability—such as the rolling of dice and drawing of lots (cleromancy), the reading of tarot cards (cartomancy), and the interpretation of dreams (oneiromancy)—have maintained their presence throughout the centuries mainly by means of fortune-telling booths and tables. Additional prognosticative arts that served as staples of fortune tellers include palm reading (chiromancy) and examining heads for telltale bumps and other signs (phrenology). Coming into prominence in the nineteenth century, the latter is one of the most recent of such methods.

As long has there have been audiences for their prognostications, fortune tellers have found seemingly inexhaustible ways to practice their craft. From sacrificing goats to reading tea leaves, from tapping for water with divining rods to measuring skulls with metal probes, practically no aspect of nature has escaped utility as a prophetic art.

Divination's Doubters

The greatest record of ancient divinatory practices was written by one of the most prominent Roman statesmen and philosophers. *De Divinatione* (Of Divination), by Marcus Tullius Cicero, likely written around 44 B.C., provides an invaluable account of the multitude of methods for foretelling the future in the pre-Christian era. Set out as a dialogue between his brother Quintus (who advocates divination) and himself (who is cynical), it captures the debate between mystical believers and hard-nosed scientists in an unprecedented manner.

Perhaps Cicero's interest in attempts to gauge the winds of fortune derives from his own inability to forecast the sharp gusts that toppled him from power in midcareer. Born in 106 B.C., he rose to fame in the time of the Roman Republic as an orator and statesman of exceptional talents. When, in 63 B.C., he risked his life and defended the Senate against armed conspirators, he was hailed as a hero. Sadly, his official acclaim proved short-lived. By hitching his career to the fate of Republican principles, his fortune was whisked away in the wake of Julius

Caesar's sudden ascension. He vied with Caesar's men, and found himself briefly in exile.

Returning to Rome, Cicero grudgingly accepted the end of the Republic and agreed to align himself with Caesar and his allies. Largely withdrawing from public life, he turned to philosophical writing as a relatively safe way of expressing his strong feelings. His keen mind and broad knowledge of Greek tradition (he had spent time in Athens and studied with disciples of the New Academy, an offshoot of Plato's Academy) led him to produce a host of important treatises, with topics ranging from politics and ethics to fate and providence.

Just when Cicero's life started to become stable again, Caesar was assassinated. Obsessed by the idea of whether or not such events could be anticipated, and whether or not it would matter if they could, Cicero wrote De Divinatione. Later, he tried to align himself with Caesar's adopted son Octavian (who became Augustus Caesar)—an alliance that eventually failed because of mutual misunderstandings. Ultimately, Cicero was assassinated by his political enemies (associated with the new ruling triumvirate). He was simply too brutally honest to have succeeded at politics, especially during those bleak days of belligerence, backstabbing, and intrigue.

De Divinatione begins with Quintus's arguments in favor of divination, which he defines as "the foreknowledge and foretelling of events that happen by chance." He divides divination into two categories: artificial and natural. The former consists of observation of auguries and portents, and includes astrology, examination of entrails, interpretation of omens such as thunder and lightning, and so forth. The latter entails prophecies uttered in altered states—through dreams and madness, including illness-based delirium. The rantings of Cassandra and the mutterings of the Pythia fall into this second category.

Quintus does not explain the exact mechanism by which divination works. Rather, he defers to the judgment of the ancients, stating that because great philosophers such as Pythagoras and Plato believed in the validity of prophecy, the notion of predicting the future by such means likely contains some merit. How could such innovative thinkers, he asks rhetorically, be completely off the mark about such an important subject? Conceding that certain purported forms of prognostication are blatantly ridiculous—snake charming, for instance—he argues that other methods based on careful reading of di-

FIGURE 1.2   *The death of Marcus Tullius Cicero, Roman orator, philosopher, and statesman, represented a ghastly day for scholarly truth (courtesy of the Library of Congress).*

vine portents could in some cases yield insight about the shape of the future.

Successful prophesies, states Quintus, rely on one or more of a few basic strategies. The first method involves coming closer to God and sensing his will. Because God presumably knows everything, by connecting with him one might catch a glimpse of the future. Deep sleep, when the mind is relaxed and thereby open to outside influence, is the state most suited to such insights into the divine. A related technique involves communing with nature, to sense its patterns and foresee its potentials. Through riding the crest of nature's waves, and sensing when its flow is most turbulent, one might anticipate its coming storms.

Another prophetic approach, according to Quintus, involves an understanding of the mechanisms of fate, which he defines as the principle that every cause creates an effect. By examining the nature of causes, one might well predict their effects. These causes have signs; so a sage well versed in the meaning of today's signals might furnish pre-science of tomorrow's events.

The second part of *De Divinatione* consists of Marcus Cicero's spir-ited rebuttal to his brother's position. He begins by distinguishing div-ination from knowledge directly obtained by the senses. Who might better interpret sensory data, he asks, a soothsayer, or an expert in the field? To him, the answer is obvious; why turn to fortune tellers, when scientific authorities generally provide far more satisfactory answers? For example, one well versed in medicine (even the rudimentary med-icine of Roman times) would diagnose someone better than would a seer with no special medical skills. Only things that expertise cannot decide should be left to diviners, he concludes.

Cicero then broadens this argument by attacking even the need for divination. Suppose one might sense the future through extraordinary means. This would only be possible if the script of tomorrow has al-ready been written. If one's fate were already sealed, then knowing it would at best be redundant, and at worst, make one absolutely miser-able.

"What do we think of Caesar?" Cicero asks.

> Had he foreseen that in the Senate, chosen in most part by him-self, . . . and in the presence of many of his own centurions, he would be put to death by most noble citizens, some of whom owed all that they had to him, and that he would fall so low an estate that no friend—no, not even a slave—would approach his dead body, in what agony of soul would he have spent his life.[5]

In other words, if Caesar foresaw the ghastly, humiliating fate in store for him, he would likely have spent his prime years moping around in a state of utter depression. There would have been no incen-tive for him to cross the Rubicon, if he knew that by doing so he was simply participating in a drama doomed to end all too tragically.

Suppose, on the other hand, the future is not yet penned, and is subject to the whims of chance. In that case, Cicero argues that divina-tion really doesn't exist. Any attempt to guess what would happen in the times to come would likely lead to changes in the present. These alterations would consequently result in abrupt changes to the world tomorrow, enough to render the initial predictions essentially moot. Deriding his brother's idea of fate, he writes:

> If it is impossible to foresee things that happen by chance because they are uncertain, there is no such thing as divination; if, on the contrary, they can be foreseen because they are preordained by fate, still there is no such thing as divination, which, by [Quintus's] definition, deals with things that happen by chance.[6]

In the final sections of his treatise, Cicero systematically debunks the main types of divination of his day—anticipating by two millennia the role James Randi (the author and magician famous for exposing fraudulent psychic claims) and other demystifiers of pseudoscientific methods play in modern times. "Divination," Cicero drolly remarks, "is compounded of a little error, a little superstition, and a good deal of fraud."[7]

One by one, Cicero pokes at the flimsy foundations of each purported method of prophecy. Astrology, he asserts, is a pointless exercise because there is no reason to think that the influence of the stars and planets—so remote from Earth—has a significant effect on newborns—or anyone else, for that matter. If the heavens affect one at birth, he argues, then why the distant celestial orbs, and not more immediate phenomena such as the weather? One hardly distinguishes between babies born on sunny versus cloudy days; why, then, discriminate between those arriving at times in which Mars is the most prominent planet, versus times in which Venus is more noticeable?

Cicero similarly mocks those who believe that dreams and omens foretell the future. Why would God choose these mysterious ways and strange times to offer his warnings, when he might do so far more directly through unmistakable messages delivered during the light of day? If a message about the future is important, and God is omnipotent, then surely he would convey it directly to its intended audience, not enshroud it in mystery.

Cicero further points out that the great majority of dreams are clearly nonsensical, and very few might be construed as possibly prophetic. Not even diehards would argue that all dreams predict the

future. Then why would God scatter a few isles of meaning throughout a vast sea of nonsense?

Meteorological and other natural phenomena would likewise be
poor media for divine messages, according to Cicero. The vast majority of the time when lightning flashes through the sky, or a cat crosses
one's path, these events have wholly prosaic explanations. Bolts sometimes fly through the air during severe storms, and no one takes heed.
Cats sometimes scramble across a road to catch mice, and no one sees
this as anything out of the ordinary. Then why, Cicero asks, would
God occasionally laden such ordinary, natural occurrences with warnings about the future? How could he expect mortals to distinguish between the commonplace and the truly significant? Most likely, Cicero
concludes, there are no such things as omens, and divination is a sham.

Cicero's treatise on prophecy apparently did not have much effect
during his lifetime. Soothsayers and fortune tellers continued to practice their craft, counseling high officials eager to learn what fate had in
store for them. Nothing in Greco-Roman worship precluded seeking
out mystical knowledge; if anything, quite to the contrary, it was encouraged for people entrusted with decision making. By Cicero's day,
the Delphic oracle had long been in decline, but seekers of prophetic
wisdom found no shortage of people eager to dispense it to them.

History proved that Cicero's words would have a much greater
posthumous impact. Within decades after his death, Christianity
emerged, and within centuries had established itself as the state religion throughout the Roman Empire. The nascent Church found itself
faced with the task of constructing a stable, social order, dedicated to
its interpretations of the teachings of Christ. Church leaders considered mystical and scientific inquiries about the future twin threats to
their mission. Jesus and his disciples had already proclaimed what
would happen in kingdom come, they argued. Therefore, they considered turning to oracles, fortune tellers, or secular natural philosophers
for additional information about human destiny redundant at best and
most often heretical.

One of the most influential of the early Christian philosophers, St.
Augustine, Bishop of Hippo, found deep inspiration in Cicero's writings. Born to a Christian mother and a pagan father in a Roman
province of North Africa in 354, Augustine had begun his lifelong
quest for truth as a follower of a mystical Persian religion,

Manichaeism. When later, after a number of long, hard years, he turned to his mother's faith for solace, he brought to Christian philosophy the experience of an outsider well versed in alternative religious approaches. He saw it as his mission to fortify emerging Catholicism with a bulwark of intelligent rebuttal to opposing views. The writings of Cicero, whom he greatly admired, helped him cement his edifice of churchly doctrines, protecting it from the volleys hurled by opposing schools of thought (including those who upheld the tradition of seeking personal mystic knowledge of the future).

In monumental works such as *City of God*, Augustine mapped out what was to become the standard Church attitude toward divination. Conceding that the Bible documents many instances of prophecy in the years before the coming of Jesus, he argued that in the Christian world prediction no longer assumed a critical role; in fact, it was counterproductive. Because Christ's appearance and resurrection constituted the culmination of history, there was no more need to look toward the future, save awaiting Christ's promised return. Furthermore, he urged, the best way to prepare for the Second Coming was to devote one's life to building the Church—not to calculate and recalculate (as many zealously did in those days) the precise date when he would reappear.[8] Hence, one should concentrate one's energies on immediate Christian tasks and eschew the temptation to waste one's time speculating about the times ahead. The details of God's future plans, Augustine concluded, should be entrusted to God, not second-guessed by mortals.

Cracking God's Code

Throughout the darkest days in the European heartland for free religious and scientific expression, scattered candles of creative dissidence still shone. In peripheral regions of Europe—Moorish Spain, southern France, and later Germany, Poland, Lithuania, and Bohemia—as well as in the Arab world (Egypt, Palestine, and so on), the flames of transcendental belief lit during Greco-Roman times continued to burn bright. Zealous believers seeking shortcuts to divine knowledge—whether through faith, science, or both—managed to evade official admonitions against prediction. Motivated by the desire to foretell the future through extraordinary means, enclaves of mystical devotees engaged in fiery debate. Each fervently hoped to be the first to decipher the thoughts of God and unravel the true secrets of the cosmos. In this manner these thinkers—mainly Jewish and Arabic—continued the quest of Pythagoras, Plato, and Aristotle, at a time in which mainstream Christians (the few who were literate) were admonished to focus solely on biblical writings.

In the great body of medieval literature known as the Kabbalah (deriving from the Hebrew "that which is received"), Jewish mystics hoped to unravel the secrets of the cosmos through numerology. Probing the Torah (Hebrew Bible) for mathematical patterns—in a process known as *Gematria*—they attempted to discover hidden knowledge about human destiny. Believing that the Torah encoded all possible aspects of the universe, they hoped to decipher its concealed messages.

The study of *Gematria* involves an enumeration of the letters of the Hebrew alphabet in which each is assigned a particular value. Applying this code, each word, phrase, and passage of Scripture are scrutinized for their numerical properties. If the sum of the numbers corresponding to one section of the Torah is equal to those of another, perhaps, according to Kabbalistic thought, they form a deep connection. As in the Pythagorean system, certain numbers harbor special meanings, relevant to the interpretation of particular texts.

The Bible Code by Michael Drosnin, a best-selling book from our own time, constitutes a modern-day attempt to treat *Gematria* in a scientific fashion. Drosnin purports that the letters of the Bible, when arranged in the proper fashion, contain hidden messages that can be gleaned through statistical analysis—particularly through correlations of related words. Within the texts of the Torah and other parts of Scripture, he argues, God has encrypted messages referring to events in the past, present, and future. According to Drosnin, premonitions of the election of Clinton, descriptions of the writings of Shakespeare, and warnings of dire moments of history, such as assassinations of the Kennedys and the Holocaust, are omnipresent in the Bible—manifest to those who understand God's code.

Drosnin particularly notes an example that seems to warn of the assassination of Prime Minister Yitzhak Rabin. If one searches in the Bible for the letters of Rabin's name, distributed in the shortest possible Equidistant Letter Sequence (ELS), one finds the expression "assassin that will assassinate" in the same passage. Apparently, he had

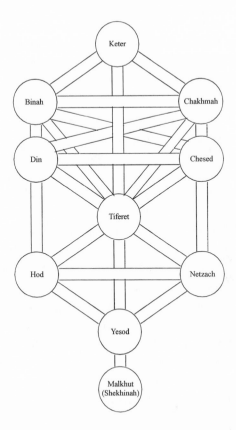

FIGURE 1.3   The "Tree of Life" formed by the ten Sephiroth is one of
the most ancient emblems of Kabbalistic Judaism. It represents the channels through which God's energy emanates into his worldly domain. Starting with the highest aspect, Keter, the crown, these powers take on various
manifestations until they ultimately reach Malkhut, the kingdom—also
known as Shekhinah, divine presence. According to Kabbalistic teachings,
each of the Sephiroth also corresponds to one of the names of God.

discovered this case one year before Rabin was gunned down and had
sent him a warning letter. The fact that Rabin had subsequently been
killed shook up Drosnin and cemented his belief that the "Bible Code"
is real.

An ELS is a distribution in which the letters of a word are equidis-
tantly spaced among the other letters of a text. A simple example of
this is the following:

"Roger O'Hara says he's Irish."

In the above phase, the word "roses" appears as an ELS with a spacing of 5. That is, ignoring spaces and punctuation, letters 1, 6, 11, 16, and 21 form "R-O-S-E-S." Indeed, this spacing within the text is the shortest ELS for that word.

The ELS that Drosnin discovered within the Torah, formed of the Hebrew letters for Yitzhak Rabin, possesses a spacing of 4,772. In his book, he depicts this by writing out the letters of the Bible, in lines of 4,772 each, and drawing a vertical rectangle around those—one above the other—that spell out the name of the late Israeli leader. He then shows how these letters intersect, in crossword puzzle–like fashion, the Hebrew expression that means "assassin who will assassinate." He attributes his discovery to God's deliberate concealment of these words in the Scriptures.

Later in the book, however, Drosnin points out predictions that he made using the Bible Code that were never fulfilled. He mentions a warning that he had given to Prime Minister Shimon Peres, Rabin's successor, that Israel would soon be the subject of nuclear attack. This event never happened, leading Drosnin to ponder that his warning might have changed history, and thus nullified what originally was meant to be.

One is reminded, by this incident, of Cicero's comments about divination. If Drosnin can change fate through his own predictions, then how—as Cicero would ask—might we conclude that the original disaster was prophesized correctly? In fact, the correct outcome—that nuclear missiles did not rain down upon Israel in the 1990s—was not foreseen in the code.

Skeptics within the statistical community, and other critics of the Bible Code point to the dangers of attributing undue significance to chance occurrences. The Hebrew language, they point out, lends itself to misinterpretation, harboring multiple meanings for each word. So if one is looking to prove a particular point by seeking semantic connections between words spaced at given intervals, through stretching meaning and loose construction, one is virtually certain to find it.

In the case of *The Bible Code*, Drosnin took considerable liberty in selecting the word pairs that formed his predictions. First, he chose only pairings that in hindsight worked out to make interesting predictions. If "Prime Minister Rabin" had turned out a better match than "Yitzhak Rabin," he reserved the right to switch from the latter to the former expression in his analysis; in another section, he used "Clinton," not "William Clinton," for example.

With great freedom of choice, each word that he selected might be written backward, forward, right side up, upside down, or diagonal. Moreover, any number of "filler letters" might appear between the letters of the word. These liberal criteria greatly increased the chances of a coincidental pairing of words forming a fulfilled prophecy. This, combined with the flexibility of Hebrew, made it likely that chance correlations might masquerade as a secret message from God.

Brendan McKay, a professor of mathematics at Australian National University and leading critic of the Bible Code, has replicated Drosnin's feat with other texts translated into Hebrew. Probing the Hebrew version of *War and Peace* with the same methods used by Drosnin, McKay found dozens of words related to the Jewish festival of Chanukah. Either these too represent cryptic messages from God, he says, or else the whole method is patently ridiculous. In this manner, he has revealed the statistical fallacies that permeate Drosnin's claim.

We have seen throughout history that chance occurrences and vague language are often distilled, purified, and siphoned out as prophecy. To bring predictions retrospectively in line with historical circumstances, loosely construed phrases are often shaken and stirred to reveal seemingly prescient pronouncements. In this manner, virtually any text might be sifted to yield a residue of purportedly fulfilled prognostications.

In the West, the Delphic priests pioneered the art of purposeful vagueness, transforming the Pythia's babbling statements into eloquent, but obscure, verse. By the time a question was presented and the answer returned to the inquirer, the information had passed through a number of filters, often rendering the response essentially meaningless. This provided the questioner with the option of reading into the reply whatever message or moral he or she wished to derive.

Similarly cryptic divinatory practices have constituted long-lasting traditions in the East. In China, for well over 2,000 years, fortunes have been told using the I Ching or Book of Changes. In this method, an inquirer throws a series of yarrow sticks or coins. Based on the results of the tosses, a special hexagram is selected from a series of sixty-four possible diagrams. Each diagram corresponds to a short text advising the questioner what his proper course should be. As in the

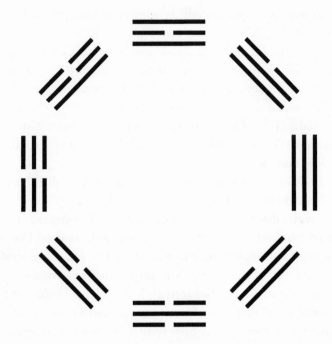

FIGURE 1.4   *Depicted is the famous Pa Kua (Eight Trigrams) emblem of the original I Ching, or Book of Changes. According to Chinese tradition, this set of symbols—a circular pattern of straight and broken line segments arranged in groups of three—was devised by the legendary Emperor Fu Hsi, who was said to have also invented writing, fishing, and trapping. For millennia, these trigrams, along with a set of sixty-four hexagrams, have been used in attempts to divine the future.*

case of the Pythia's pronouncements or the predictions of a Kabbalist, each passage is sufficiently nebulous as to be open to many possible interpretations. "Act with righteousness," is a typical example of what the I Ching's response might be to a request for advice on the proper action to take.

One of the quintessential examples of vagueness in prophecy was provided by Nostradamus in his book of predictions. For years after his death, interpreters have read what they wanted to read into his elusive verse. Thus, by refusing to be specific, he maximized the lasting power of his statements.

The Eyes of Nostradamus

Michel de Nostredame, better known as Nostradamus, was born in 1503 in the Provence region of southern France—a leading center of mysticism—to a family of recent Jewish converts to Catholicism. As a youth, his grandfather, a doctor and herbalist, taught him mathematics and astronomy, as well as Greek and Latin. Following the path that his revered grandfather had trod, he studied medicine at Montpelier University, graduating in 1525.

As a physician, Nostradamus was said to have been brave and resourceful, dealing especially effectively with victims of the plague. At the time, many doctors flinched when it came time to treat plague victims, worried that they would become infected as well. Other physicians thought that bleeding patients would help drain their pestilence and cure them. Nostradamus took a third course, neither fleeing nor taking hasty measures, but, rather, calmly attending the afflicted while insisting on proper hygiene and the use of antiseptics. As one would imagine, his methods led to much greater success in treatment.

Later in life, Nostradamus spent more and more time alone, in a state of deep contemplation. He became convinced that he possessed clairvoyant powers, and felt that he could summon these when he was in relaxed states of mind. Staring night after night at a water-filled vessel mounted on a brass tripod, he meditated on the shape of the times to come. His mind became engaged with ghastly visions of warfare, famine, disease, fire, floods, and other calamities.

Following the procedure used by the priests of Delphi, Nostradamus wrote down his thoughts as poetry, portraying each image of the future in four lines of verse. The language he used to capture his visions on paper was far from straightforward. Each quatrain (four-line poem) plays out in an elusive, tangled fashion, full of cryptic references, obscure metaphors, strange anagrams, and even phrases from foreign tongues.

In 1555, Nostradamus published the first collection of his quatrains, which he called *Centuries*. The French term "centurie" has nothing to do with time, but rather refers to the organization of the book into volumes of 100 verses each. When his project ended with his death, slightly more than a decade later, he had completed almost 1,000 quatrains, grouped into ten volumes. His life's work was first published posthumously in 1568, and has been reprinted throughout the years in countless versions, translations, and interpretations.

One curious aspect of Nostradamus's collection of prophecies is that they appear completely out of chronological order. In any given sequence, one might refer to a kingly succession during the Renaissance, the next to a tragedy in the twentieth century, and the next to a revolution taking place sometime in between. Some scholars have argued that Nostradamus deliberately jumbled the historical progression—for the same reason he used obscure language—to diminish the possibility of offending those to whom he refers and to confuse potential critics, including the powerful Inquisition. Other historians believe that Nostradamus was looking ahead to future audiences, and realized that in different ages varying interpretations would be made, guaranteeing his work's longevity. Yet others attribute the haphazard nature of the centuries purely to the random quality of prophetic visions (who has ever had orderly dreams?), and acquit its author of the charge of deliberate obscurity.

Nostradamus does made it explicitly clear, however, in a letter to his son included in the 1568 edition, that his clairvoyance extends to, but not beyond, the year 3797. Curiously, though, there are very few dates mentioned in the quatrains, and none beyond the year 1999, when Nostradamus ominously predicts "the great king of terror will come from the sky." That year has come and gone, and still nobody knows what that statement means.

Aside, perhaps, from the frightening prediction for 1999, the most widely interpreted passages of Nostradamus's prophecies have been taken to refer to Napoleon and Hitler. One quatrain beginning with the line, "From being a simple soldier, he shall come to rule an empire" has been applied first to Napoleon, then later to Hitler. Note that this passage could refer to almost any conqueror; practically any general has once been a simple soldier.

Other verses, referring to a "Hister," have been pageanted by devotees as proof that Nostradamus anticipated the rise of the German führer. This claim seems quite dubious, however, when one realizes that—as renowned magician, skeptic, and MacArthur fellow James Randi points out—one of the ancient names for the Danube is "Hister."

During World War II, the Nazis, particular Goebbels, were fascinated by some interpretations that seemed to prophesize when and how they would come to power, and dismayed by others that seemed to forecast they would lose the war. Consequently, Goebbels, who was minister of propaganda, commissioned faked Nostradamus predictions favorable to the Nazis, and broadcast them to the Allies as part of his psychological warfare campaign. Needless to say, they had little impact on Allied morale.

Today, the meaning of the *Centuries* remains the subject of much contention. Most scientists, historians, and other scholars—including Randi and other members of the "skeptical community"—assign little credence to the notion that Nostradamus, or anyone else, could have supernatural visions of the future. Pointing out the vagueness of statements by Nostradamus, as well as claims by more recent self-proclaimed prophets, such as Edgar Cayce and Jeane Dixon, Randi has devoted considerable energy debunking those whom he calls "people who peddle nonsense."[9] Nevertheless, much to Randi's chagrin, Nostradamus, almost half a millennium after his death, maintains a bevy of enthusiastic followers (many of whom also follow Cayce and Dixon).

During his lifetime, Nostradamus did not have to face such controversies. Only toward the end of his life did the Justices of Paris begin to inquire about his practices, and by then he was well protected by his regal connections; his patron was Catherine, Queen of France. When Catherine's son, King Charles IX, succeeded her husband, King Henri II, to the French throne in Arles, Nostradamus was retained as royal adviser, and awarded the handsome sum of 300 gold crowns. He died rich, but relatively obscure; the fame and controversy would come much later.

We have journeyed a long way in history, from the Temple of Apollo to the court of Arles. Yet it is remarkable how much these two periods had in common, in terms of the relationship between mysticism and the state. Apparently, translating visions into verse and passing that off as policy has been a good way of influencing crowned heads—and gaining rich rewards from them to boot.

2

THE LATHE OF LAPLACE

The Deterministic Future

*He did read the riddle of the heavens. And he
believed that by the same powers of his
introspective imagination he would read the
riddle of the Godhead, the riddle of past and
future events divinely foreordained, the riddle
of the elements and their constitution from an
undifferentiated first matter, the riddle of
health and immortality. All would be revealed
to him if he could only persevere to the end.*

—JOHN MAYNARD KEYNES
Newton, The Man

Season of Change

The close relationship between mysticism and science finally began to
unravel toward the end of the sixteenth century. Until that point, public perception—as well as the tenets of state and religious bodies—held
that mystical divination and the scientific practices of prediction were
virtually one and the same. From the time of Pythagoras until the age
of Nostradamus, the art of foreseeing the future was perceived as pagan
(black) magic—seen as virtuous or evil, depending on the mood of the

day. "Sorcerers" and scientists shared the same heritage and fate—venerated in auspicious times and threatened with burning at the stake when times grew sour.

By the start of the seventeenth century, the paths of occult practitioners and scientists were beginning to diverge. Although much of the mystical community still found itself in mortal danger for its beliefs—witchcraft trials were relatively common during that era—scientific thinkers began to enjoy freer discourse, heightened respect, improved access to ancient and contemporary texts, and a healthy sense of progress. This was particularly true in the northern nations of Europe, where new secular universities, such as the University of Leiden in Holland and the University of Edinburgh in Scotland opened their gates to scholars of all faiths and all philosophies. Because of Church conservatism, it would be a while longer before academicians enjoyed the same freedoms in the south—still, radical philosophers such as Giordano Bruno pushed forward the boundaries of freedom with expanded views of the universe, risking or even sacrificing their lives (as Bruno did) in the process. These ventures eventually led to an improved intellectual climate in the southern nations as well.

At the height of the great Age of Exploration, Europe had entered an era of unprecedented scientific achievement. In parallel with the discovery and settlement of hitherto unexplored (by Europeans) regions of the globe, novel aspects of the natural world were being mapped out. The distribution of two revolutionary texts, both published in 1543—*On the Revolutions of Celestial Bodies* by Nicholas Copernicus and *On the Structure of the Human Body* by Andreas Vesalius—had unleashed a torrent of discussions about the nature of outer and inner space.

Elizabeth I had acceded to the throne of England and was soon to launch the battles that led in 1588 to the defeat of the Spanish Armada. Political and economic power was shifting northward, where religious authorities yielded less power over intellectual life. Shakespeare's plays (written around the turn of the seventeenth century) were captivating hearts, expanding minds and enriching the English language, and Francis Bacon's writings, including "Of Innovations" (1601), *The Advancement of Learning* (1605), *Wisdom of the Ancients* (1609), *Novum Organum* (1620), and *The New Atlantis* (1624) were scoping out bold new visions of science. Bacon's portrayal of science as the new frontier, with scientists as leaders and explorers, helped rally unprecedented public enthusiasm for the field. Moreover, his advocacy of objective collection and verification of data launched the "scientific method"—the basic procedure for experimentation that is still used today.

Even before Bacon published his treatises, experimental researchers began to challenge the long-held assumptions of the Greeks. Emboldened by the newfound liberties of the times, they reexamined classical paradigms of nature, making minor corrections or radical changes, as need be, when new data did not seem to fit old models.

In 1586 Flemish mathematician Simon Stevin demonstrated that Aristotle's theory of mechanics contained a major error. Contradicting Aristotle's notion that heavier physical bodies drop faster, Stevin showed that objects fall to the ground at speeds independent of their weight. Three years later Galileo independently, and more famously, reached the same conclusion.

About the same time, Stevin proposed sweeping reforms in mathematics, introducing (or in some cases popularizing) real numbers, positive, negative, and square root symbols, and the modern system of decimal notation. By incorporating these innovations into his published texts, Stevin demonstrated their utility for all technical endeavors. Indeed, without these new symbols and methods, it's hard to imagine how modern science could have launched itself beyond the narrow perimeters of classical knowledge.

Meanwhile, Tycho Brahe of Denmark, possibly the greatest naked-eye astronomer of all time, embarked on a multidecade quest to modify and expand the ancient visage of the heavens. His observational data, along with Johannes Kepler's ingenious interpretation, changed the course of predictive science forever. By presenting Kepler with a precise set of information about planetary motions, Tycho enabled him to develop laws of orbital behavior. This was the first time in history a physical forecast, rendered in exact mathematical terms, accurately anticipated the movements of objects. Thus, the work of Tycho and Kepler served as a prototype for the predictive machinery of classical physics.

Born in 1546 to a noble family, Tycho was extremely headstrong and haughty, even as a youth. He had a strange appearance, resulting from a row in his student days. While fighting a duel with another boy over who was better at mathematics, the bridge of his nose was sliced off. The missing part was promptly replaced with a boxy gold and silver piece. Perched on top of a handlebar mustache, his unusual proboscis only added to his regal mystique and air of authority.

At the University of Copenhagen and then at several German universities, Tycho took up the study of astronomy. Learning the skills of the trade, he was mortified by the sloppy techniques used for mapping celestial movements. Data was collected only sporadically, mainly to supplement ancient observations. In most cases, centuries-old readings in withered texts had never been verified. Vowing to improve the field dramatically, he dedicated his life to measuring astronomical phenomena with pinpoint accuracy.

In 1572, after completing his education at the age of twenty-six and returning to his native land, Tycho made the phenomenal discovery of a bright new star in the heavens. Up to that point, influential religious doctrines held that the stellar firmament was essentially static, having been created at a single time in the past. Even though other new stars had been seen from time to time (the one recorded in the story of Christ's birth, for example), they were considered miracles or omens and assigned no astronomical significance. Tycho's spectacular object, in contrast, was the first to be scientifically analyzed and classified as a new-sprung astral phenomenon. Thus, by demonstrating that the celestial map is hardly frozen in time, but rather is subject to sudden changes, even explosions, the radiant visage of Tycho's "nova" symbolically ushered in a spirit of change in astronomy.

The Danish crown rewarded Tycho's diligence by presenting him with his own custom-built observatory, a colossal fortresslike structure perched on the island of Hveen. Within the confines of his own domain, Tycho found himself king of his island world by day and master of the heavens by night. Every evening he slowly and painstakingly plotted out the positions of the visible planets relative to the "fixed" stars, and jotted down his measurements in one of an endless series of notebooks. Then, by day, he would eat heartily, drink excessively, and strive to impress his many guests—that is, when he wasn't busy berating his servants.

Tycho's drunken demeanor and explosive temper soon proved too much for the people of Hveen. As disdain for his antics rose to a crescendo, he soon got the idea and decided to move elsewhere. In a

*FIGURE 2.1   The tomb of Tycho Brahe
(1546–1601), perhaps the greatest naked-
eye observational astronomer (courtesy of the
Library of Congress).*

stroke of luck, a new position opened up for him in Prague, Imperial
Mathematician to Emperor Rudolph II. Granted a handsome salary
and the castle of his choice to use as a residence and observatory, Tycho
settled down in Benatek Castle, where he commenced his methodical
charting of planetary trajectories. Soon his voluminous sets of notes
were even further packed with new astronomical data.

In mapping out the solar system, Tycho maintained an omnipresent
goal: to develop a novel and accurate theory of planetary motion. Previous attempts to do so were either highly inaccurate—Aristotle's
model of simple concentric circular motion around Earth—or extraordinarily complex—Ptolemy's theory of epicycles, in which planets

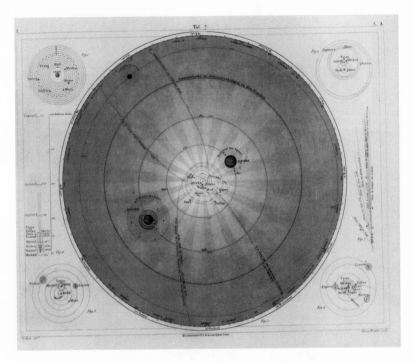

FIGURE 2.2   This early diagram depicts the relative positions, sizes, and orbits of the six planets known in the days of Johannes Kepler and Tycho Brahe. It also shows Uranus, the seventh planet, discovered by William Herschel in 1781 (courtesy of the Library of Congress).

dance in small circles as they complete large revolutions around Earth. Even Copernicus, who in the beginning of his groundbreaking treatise correctly pointed out that the planets revolve around the Sun, not Earth, concluded his proposal with a more intricate theory.

What was the dilemma? Why couldn't these great minds accurately model something as basic as the movements of the planetary bodies? The problem had to do with the phenomenon of retrograde motion. Certain planets appear at times to move backward in the sky relative to their usual motion, and then to turn tail once more and resume their normal course. Assuming that the planets move in circles around Earth (or even around the Sun) the only way to account for these patterns is to add little circles, or epicycles, to the big circles.

FIGURE 2.3 Johannes Kepler
(1571–1630) (courtesy of the AIP
Emilio Segre Visual Archives).

Tycho wished to do better. He was certain that he possessed the best data, but was not sure if he knew what it all meant. Unfortunately, he lacked the mathematical know-how to transform his detailed observations into a successful new cosmology. Not that he didn't try. Afraid of being bested by competing astronomers, he secretly tinkered with variations of Ptolemy's model until he was blue in the face (or at least a little less red), trying to reduce the number of complications. He finally advanced his own theory—which scarcely seemed better than the others.

It is doubtful that the lyricist Ira Gershwin would have been quite as famous if it weren't for the gifted ear for music of his brother George. Although Ira had a knack for clever lines, George's infectious tunes, heartfelt strains, and bold rhythms brought these ideas to life. What Tycho needed was a genius to round out the coarse data of his comprehensive survey of the solar system. When Johannes Kepler contacted him requesting to visit his observatory, he felt lucky indeed. As Ira versified such feelings: who could ask for anything more?

The Geometry of Perfection

Four hundred years ago an extraordinarily unlikely meeting between two extraordinarily different people held extraordinary implications for the science of prediction. Two refugees—one from religious oppression, the other from the effects of his own fiery temper—found themselves working in the region of Prague at the same time. Kepler, a German Protestant from a poor family, was as somber and serious an introvert as they come. A deeply religious man, trained as a theologian, he had just lost his mathematics teaching position in the Austrian Catholic town of Graz because of his beliefs. Tycho—the very model of wealthy flamboyance—had practically been booted out of Denmark because of his loud mouth and foul habits. Tycho's arrival virtually coincided with Kepler's need for a respectable position, which the Dane heartily provided to the young German. Thus, a collaboration that proved monumentally important for astronomy, physics, and science in general came about largely by chance; they probably never would have met if it weren't for their respective "banishments."

It's clear why Kepler left Graz. But why did he hone in on Tycho's observatory in particular? Kepler's main motive for choosing Benatek Castle instead of other possibilities was to gain access to what he saw as the best set of astronomical data available. He needed this information to help him resolve burning questions that he had about the cosmos.

As a teacher of mathematics in Graz, Kepler had been fascinated by the notions of form and symmetry that permeated Euclidean geometry. When he was not instructing students or moonlighting by selling homespun horoscopes (a common sideline at the time for those who knew something about astronomy), he meditated deeply on the elegance of shapes and patterns, particularly the exquisite completeness of the set of forms known as the Platonic solids.

The Platonic solids, whose discovery has been attributed to either Pythagoras or Plato, comprise the set of all possible regular (equal-sided) polyhedra. Polyhedra are multisided three-dimensional objects; that is, geometric solids. Regular polyhedra (four-sided pyramids, cubes, and so on) form higher dimensional generalizations of regular polygons (such as equilateral triangles and squares). Curiously, though, although there are an infinite number of types of regular polygons, there is only a small set of regular polyhedra—precisely five in all.

Why nature—usually bountiful—counts these basic figures on only one hand has puzzled philosophers throughout the ages.

Another numerical mystery—with which Kepler saw a profound connection—concerned the presence of only six planets (known in his day) in the solar system. Why just six, and not hundreds or thousands? he wondered. Perhaps, he pondered in an exhilarating moment of inspiration, the arrangement of planets in the heavens relates in some essential fashion to a special ordering of the Platonic solids.

Like Pythagoras, the Kabbalists, and other mystical figures before him, Kepler felt that he had unearthed profound mathematical truths about the universe. He had come to believe that God, in his infinite eloquence, had expressed his deepest secrets both in the theorems of geometry and in the laws of the cosmos. Moreover, in his economy of forms, he had used the same basic principles to guide both. Knowledge of this code, Kepler supposed, would eventually lead to an understanding of the nature of human destiny. Though he realized that astrology as it was then practiced was far too simplistic, Kepler thought that science would eventually unravel connections between astral harmonies and mortal fate. His discovery, he felt, was the first step along that sacred path of knowledge.

Unlike Tycho, Kepler was a firm believer in the Copernican notion that the planets revolve around the Sun, not Earth. Therefore, in his attempts to describe the positions and paths of the planetary bodies, he proposed that the Sun would play a central role, and imagined a series of concentric Platonic solids and spheres stacked around it. Kepler placed these, in his model, one inside the other, like Russian *matryoshka* dolls.

Kepler's clever concoction of Platonic solids failed to match known planetary data, but led him to spend many arduous years studying detailed astronomical results. He finally developed a less aesthetically pleasing, but highly accurate, model that inspired Newtonian mechanics. Thus his youthful passion for mathematical elegance ultimately yielded a mature understanding of natural law.

This process of realization took many steps. At first, Kepler assumed that his geometric construct didn't match the astronomical data because the information was old, and therefore unreliable. He realized that to refine his model he needed far more accurate measurements of planetary positions. His search for such up-to-date and complete celestial knowledge set him on an inalterable course to the observatory of the greatest observational astronomer of his day. As he wrote in his diary: "Tycho possesses the best observations, and thus so-to-speak the material for the building of the new edifice; he also has collaborators and everything he could wish for. He only lacks the architect who would put all this to use according to his own design."[1]

Kepler exchanged several letters with Tycho before they met. He arrived in Prague with the hope that Tycho would supply him with all the measurements he needed, but the gilded-nosed Dane was extremely jealous with his data. Though he was quite generous with his food and drink, serving guests bountiful helpings of each, he doled out his planetary information with the magnanimity of a Dickensian orphanage.

In general, the one and a half years that Kepler spent with Tycho were extremely unpleasant. Their clashing temperaments often brought them to red-faced battle. The two great men argued and argued—Kepler tried to wrest data from Tycho, and Tycho attempted to convert Kepler to his own system. Yet Kepler, in exile from Graz with nowhere else in particular to go, clung to the hope that if he remained in Benatek Castle he would eventually get what he wanted. And he did, in time.

In August 1601, in a meeting between Kepler, Tycho, and Emperor Rudolf II, Kepler gained official recognition as Tycho's assistant. Within months, Tycho became gravely ill from complications arising from a urinary infection. After Tycho died on October 24, 1601, Kepler succeeded him as imperial mathematician. On his deathbed, according to Kepler, Tycho deliriously pleaded to his successor, over and over, "Let me not seem to have lived in vain."[2]

The new imperial mathematician wasted no time delving into his long-awaited treasure chest. At last, the finest planetary data of his day was his to access. Kepler marveled at the precision of Tycho's measurements, particularly at the detailed information gathered on the path of Mars. Tycho, along with his assistant Longomontanus, had watched Mars closely because of the prominence of seemingly erratic features in its orbit, such as eccentricity (elongation) and retrograde motion, which could not be well matched by either Ptolemaic or Copernican approaches. Kepler shared Tycho's belief that understanding why Mars

FIGURE 2.4   Mars, the planet in the inner solar system with the most eccentric orbit, represented a challenge to those trying to understand cosmic dynamics. Ultimately it was Johannes Kepler who cracked the riddle using Tycho Brahe's exemplary data (courtesy of NASA).

appeared to behave strangely would provide critical help in piecing together the puzzle of the entire solar system.

"I believe it was an act of Divine Providence that I arrived just at the time when Longomontanus was occupied with Mars. For Mars alone enables us to penetrate the secrets of astronomy which otherwise would remain forever hidden from us."[3]

Kepler tried in vain to adjust his Platonic solids approach to fit Tycho's measurements. When that didn't work, he attempted to verify the Copernican approach by describing the motion of Mars as a simple circular path around the Sun. To his chagrin, that didn't work out either. Much to Kepler's amazement, the Martian trajectory could not be modeled as a circle or set of circles very well at all; it could best be described as elliptical. Turning to Tycho's data about the other planets, he found the same effect. Apparently, Kepler discovered, all planets move in elliptical, not circular, orbits around the Sun.

When Kepler reached his conclusions about the shape of planetary movements, he blazed an entirely new approach for predictive science. He brilliantly demonstrated how a theorist might mold a set of gathered experimental data into a successful mathematical model. Putting his own prejudices aside for the sake of intellectual honesty and empirical truth, he exhibited an open-minded attitude ideal for scientific exploration. Though he followed the Pythagorean tradition and harbored strong philosophical preferences for symmetric, regular objects, such as circles and spheres, when facts led him in a different direction, he embraced the radical idea of elliptical orbits. His turnabout was akin to a Renaissance art expert, immersed all his life in Botticelli and DaVinci, encountering Picasso and suddenly becoming a leading advocate for cubism. Because of his willingness to change his perspective when experimentally gathered evidence demanded it, Kepler led the way to a new era of fruitful, rigorous scientific endeavor.

Kepler was fortunate that his data was regular enough that by making adjustments for perspective (picturing it from the Sun instead of Earth) he could model it by means of a simple geometric shape. More recently, analysts have grappled with data of far greater complexity and have been forced to use a variety of computational algorithms and statistical techniques. Often, predictions rendered from such complicated sets of information harbor little of the simplicity of a fundamental natural law. Luckily, the German mathematician inherited a relatively clean body of measurements, suitably for the basic analysis of his times—namely, recognizing geometric patterns.

Kepler's elliptical finding comprised but one of his three basic conclusions about planetary motion—a hallowed set, found in any physics textbook, known as Kepler's laws. His second and third principles prove no less revolutionary. Orbiting planets, he found, sweep out equal areas in equal times. Moreover, the squares of their times taken to go around the Sun are proportional to the cubes of their average distances from the Sun.

When Kepler wrote of planets sweeping out uniform regions of space, he had a vivid image in mind. He imagined the Sun's influence as a colossal rotating broom, whisking along objects in its path as it turns, like a robot housekeeper spinning around and methodically sweeping dirt off an oval floor. This broom possesses the power to push each planet at different rates along separate elliptical orbits.

Pythagoras pictured a central fire around which celestial orbs danced in harmonic resonance. Neglecting physical considerations for such behavior, he imagined mystical, musical, and mathematical influences drumming the cosmos into motion. Plato, less mysterious but equally metaphysical in his approach, considered the circular rhythms of heavenly bodies expressions of divine perfection. Neither man sought tangible explanations for these patterns, realistic reasons why planets would be compelled to revolve. Even Copernicus, developing his more accurate heliocentric model, apparently never took the time to ponder the driving forces spinning his system of concentric wheels.

Kepler, in contrast, possessed a keen awareness of the need to justify *why* his theory worked, not just *how* it behaved. Moreover, though he had deep religious convictions, he felt that his explanation needed to be secular in nature. He refused to simply observe nature through science, describe nature through science, but then envelop his package of observations and descriptions with supernatural wrapping paper.

In 1609, Kepler detailed his arguments in a massive treatise with the bold title, *Astronomia Nova* (New Astronomy). In it, he presented his primitive notion of forces, explaining the Sun's sweeping influence by arguing that certain objects might physically affect other objects over considerable distances. He hadn't quite encapsulated gravity—he bequeathed that project to Newton—but in pondering long-range astrophysical effects he set the groundwork for the predictive success of classical mechanics.

Hopeful that his book would spark a bonfire of altered belief about the celestial spheres, Kepler sent a copy to another leading astronomer of his day, Galileo Galilei. Galileo, a firm believer in the Copernican model, circular motion and all, did not know what to make of Kepler's suggestions, and never honored his colleague with a proper response. Instead, the renowned Italian set out to examine the universe with his own invention: the first telescope designed for astronomical use (the *very first* telescopes were assembled by the Dutch spectacle maker Johann Lippershey in 1608).

Galileo's telescopic data perfectly supplemented Kepler's hand-plotted results. Observing mountains on the Moon, phases of Venus, and satellites circling Jupiter, Galileo found visual evidence that the planets were worlds in their own right. Along with Kepler's laws, Galileo's discoveries comprised the launching pad of modern astronomy, from

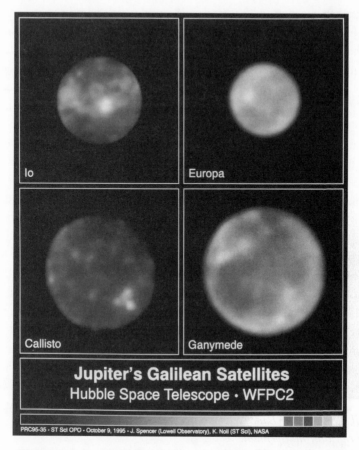

FIGURE 2.5   Detecting the four largest moons of Jupiter was
one of Galileo's great astronomical achievements. It helped
demonstrate that other planets harbor properties similar to
Earth. Galileo named these objects the "Medicean stars" after his
mentors, the Medicis, but we now know them as Io, Europa,
Callisto, and Ganymede (courtesy of NASA).

which humankind's centuries-long optical exploration of outer space
rocketed forward.

Galileo's liberal interpretation of the Bible (which Kepler shared less
vocally and less dangerously), anathema to the religious authorities of
the times, similarly propelled cosmology forward. Believing that scientific depictions in Scripture were often simplistic or even metaphorical,
he valued observational evidence far more than biblical accounts. In

FIGURE 2.6   Sir Isaac Newton (1642–1727), developed classical physics and calculus, indispensable tools for modern predictive science (courtesy of the AIP Emilio Segre Visual Archives, W. F. Meggers Collection).

his famous "Letter to the Grand Duchess Christina," he challenged theologians to refute scientific conjectures with hard facts, not rhetoric, writing: "Truly demonstrated physical conclusions need not be subordinated to biblical passages . . . Before a physical proposition is condemned it must be shown to be *not* rigorously demonstrated."[4]

The subject of astronomy was not the only beneficiary of the truths discerned by Kepler and Galileo. The greatly enriched understanding of the solar system provided by these pioneers led to an enhancement of all fields of science and a heightened impetus for mapping out every discernable feature of the natural world. The predictable motions of astral objects through space became seen as paradigms for successful forecasting in general, from calculating the paths of billiard balls to computing the rhythmic movements of pendulum. As the ancients believed, and classical scientists reiterated, those who fathom the heavens hold the key to understanding worldly events as well.

Newton's Obsessions

Isaac Newton, founder of classical physics and one of the greatest scientific geniuses of the second millennium, was born in Lincolnshire, England, in 1642, the same year that Galileo died. An enigmatic man, his personality displayed as many angles as the prisms he famously studied. There was a public Newton, a private Newton, and an even more private Newton. Publicly, Newton was esteemed, extolled, mythologized, and otherwise venerated. The first scientist ever to be officially called "Sir"(he was knighted by Queen Anne in 1705), and the first to be buried in Westminster Abbey—beneath an ornate stone monument, carved with astronomical and mathematical symbols—he was, in essence, the first scientific superstar.

Those who encountered Newton through academic pursuits, however, became familiar with his obsessive private side. Possessing profound analytic abilities, he could sit down to solve a problem and end up founding a whole field. Said to have no hobbies, during his younger years he spent virtually all his time buried in his studies. Even later in life, when he took on several public roles, including Master of the Mint and the prestigious Royal Society presidency, he remained reclusive, never marrying or establishing close friendships.

There was yet another side of Newton that was unknown, even to historians, until certain of his private papers were unearthed. In 1936, economist John Maynard Keynes, famous for his advocacy of government spending, bought a set of Newton's writings at auction. He promptly donated them to King's College, Cambridge, where Newton's other papers had been collected. Within these newly found manuscripts contained proof that Newton spent much of his time investigating alchemy, divination, and other mystic pursuits, especially prophetic passages in the Bible. Unraveling Daniel's dreams about horrific beasts contending at the end of time was one of his strongest obsessions. With his burning desire to decipher Scripture and become privy to the very thoughts of God, Newton, unbeknownst to his contemporaries, secretly possessed the soul of a Kabbalist.

A glance at Newton's personal library—which none, save his servants, had access to, would have revealed the depth of Newton's passion for the occult. Contemporary research has shown that of his 1,752 books, fully 170 had mystical content, and only 369 had scientific themes. Included in the former category were dozens of alchemical works, as well as collections of Kabbalistic and other occult movements' teachings (Hermetism, Rosicrucianism, and so forth).

In a famous essay published in 1947, Keynes asserted that Newton's passion for mysticism drove his quest for understanding the cosmos, and ultimately led to his great discoveries. In spirit, Keynes argued, Newton was more a "magician" than a scientist:

> Why do I call him a magician? Because he looked on the whole universe and all that is in it *as a riddle,* as a secret which could be read by applying pure thought to certain evidence, certain mystic clues which God had lain about the world to allow a sort of philosopher's treasure hunt to the esoteric brotherhood. He believed that these clues were to be found partly in the heavens and in the constitution of elements (and that is what gives the false suggestion of his being an experimental natural philosopher), but also partly in certain papers and traditions handed down by the brethren in an unbroken chain back to the original cryptic revelation in Babylonia. He regarded the universe as a cryptogram set by the Almighty—just as he himself wrapt the discovery of the calculus when he communicated with Leibniz. By pure thought, by concentration of mind, the riddle, he believed, would be revealed to the initiate.[5]

Hence, as in Kepler's case—perhaps even more so—Newton's impetus to resolve scientific questions about the cosmos derived from a mystic quest for the meaning of human destiny. His ardent desire to unlock God's safe and sneak a glimpse at the Book of Life provided an incentive for him even stronger than personal fame to spend virtually all his waking hours (at least in his younger days) formulating novel mathematical and physical techniques.

Naturally, given his drive for deeper cosmic understanding, Newton was greatly interested in Kepler's laws of planetary motion and Galileo's astronomical discoveries about celestial bodies. Some property of the planets and the Sun seemed to compel them to engage in a great dance spanning millions of miles. Something about the Moon caused it to waltz again and again around Earth, always facing its greater partner. What influential power persuaded each of these distant bodies to take part in the grand cotillion of the spheres?

The answer, Newton realized, is that the same force that draws falling objects on Earth down to the ground also causes mutual attraction between any pair of massive bodies in the universe. Gravity is the invisible glue that binds the solar system and prevents everything from flying apart. Newton found that the magnitude of this force decreases with the square of the distance between two objects, justifying why Mars experiences much, much less of Earth's attraction than the Moon.

To explain the movements of celestial bodies, Newton realized that he needed to do more than simply define their driving force. He also had to detail precisely how this gravitational attraction—and other types of pushes and pulls—cause motion to occur. Does any kind of motion require a continuously applied force, he wondered, or only particular movements?

His elegant answer to this question has come to be known as Newton's laws of motion. These principles are far broader in scope than Kepler's laws. They apply not only to planetary behavior, they pertain to the widest range of activities imaginable, from the collision of organic materials with cell membranes to the whirling of stars around galactic centers.

Newton's first two laws consider differences in outcome whether an external force (or set of forces) is applied to an object. In the absence of such an outside push or pull, a body is said to be in the state of inertia. If it is at rest, it remains at rest. If it is already moving, on the other hand, it keeps going in the same direction at the same speed. For this reason, if an astronaut far away from Earth throws a baseball out into the void, it will continue to move along a straight line forever (or at least until the gravity of a star or planet reigns it in). Professional league games will probably never be played in deep space; chalk that up to the law of inertia.

All this changes once an external force is applied. Newton's second law holds, in that case, that an object tends to accelerate at a rate proportional to the amount of push or pull it encounters. The more force, the more acceleration, in a ratio equal to the object's mass.

To complete the picture, Newton's third law relates to the mutual interaction between two bodies. Every action—produced by the first body and affecting the second—produces a reaction of equal magnitude but opposite direction—produced by the second body and affecting the first. For this reason, because Earth exerts a strong gravitational force on the Moon, capturing it in orbit, the Moon exerts an equal reactive force on Earth, creating the lunar tides.

Newton's principles apply in a venue called absolute space and absolute time. Absolute space assumes that all distances seem the same to all observers, like the regular spaces on a checkerboard viewed similarly by players sitting on each side of it. Absolute time implies that each second passes in the same manner for everyone under any circumstances. Though these statements might seem trivially correct, when we consider Einsteinian relativity, we'll examine their constraints.

Together, Newton's laws can be used to chart the future behavior of an object, whenever its current conditions are known exactly. Newton showed that a particle's situation in space might be wholly expressed in terms of only two properties: its position and its velocity. Combining this information with knowledge of its mass and all its experienced forces—gravitational attraction or electrical repulsion, for example—provides enough data to determine its position and velocity for all times. Knowing a pitcher's location, and the speed and direction of her throw, for instance, yields information about how high the ball she pitches will go, where it will land, and how fast it will impact upon the ground. Hence, in theory, Newtonian physics can be used to forecast any object's trajectory.

Though Newton's principles have their limitations—as we'll see when we discuss relativity and quantum theory—they render accurate predictions for an extraordinarily diverse kaleidoscope of physical phenomena. Scientists and philosophers have been captivated by their mechanical regularity. The power and range of their predictive capacities have inspired diverse visions of an eternal clockwork universe—spinning its infinite set of gears without sign of wear or tear—functioning flawlessly until the end of time.

The Clockwork Demon

The eighteenth-century French mathematician Pierre Simon Laplace, who lived at a time when Napoleon strode through Europe and defied the might of Britain, had the high ambition to conquer the greatest possible arena of physical phenomena and stake a claim (albeit a friendly one) as master of British science. Newton died in 1727, and, alas, being only human, had left many practical applications of his theory unsolved. Compared to his dazzling youth, he hadn't done as much basic science in his later years, anyway, after a nervous breakdown he had suffered in 1693.

Born in 1747, Laplace was quite a confident researcher, and saw the mountain of problems to be scaled through the laws of Newtonian physics as a climber's challenge. With great dexterity, he hoisted his way over the difficulties posed by hundreds of formidable questions, including many riddles involving the planets. For example, he resolved an issue that had puzzled Newton: why Jupiter and Saturn periodically change their orbital patterns. By clever use of Newtonian principles Laplace discovered the answer. The two planetary giants come closest to each other about once every fifty-nine years and perturb each others' orbits through their mutual gravitational attraction. He presented this solution, along with the results of his other studies, in a thick, five-volume collection, titled *Celestial Mechanics,* published between 1799 and 1825.

Not only did Laplace include calculations and numerical methods in his treatise; he also added a good measure of his own views about natural law. Laplace was extremely interested in the philosophical as well as the practical applications of Newton's work. Based on his own success with resolving seemingly intractable problems in physics, such as the Jupiter-Saturn question, he had come to the conclusion that any issue in nature might, in theory, be solved by means of Newtonian mechanics. He expressed these notions, in his work, as a tribute to Newton's genius, as well as an archetype for future science.

In an assertion that has come to be known as Laplace's demon, he claims that any being who was aware of everything about the universe at any given moment would understand everything about it for all times. As Laplace wrote:

FIGURE 2.7 Henri Poincaré (1854–
1912), founder of the modern science of
dynamical systems (courtesy of the AIP
Emilio Segre Visual Archives).

An intellect which at a given instant knew all the forces acting in
nature, and the position of all things of which the world consists—supposing the said intellect were vast enough to subject
these data to analysis—would embrace in the same formula the
motions of the greatest bodies in the universe and those of the
slightest atoms; nothing would be uncertain for it, and the future, like the past, would be present to its eyes.[6]

Laplacean determinism, as expressed in this statement, proved extremely fashionable in Europe throughout the nineteenth century. At
the dawn of the industrial age, mechanistic explanations of the work-

ings of the universe held a natural attraction for those accustomed to the incessant din of factory engines. Like the rolling wheels of a locomotive and the spinning gears of a turbine, ceaseless, rhythmic activity seemed a natural way of characterizing reality itself.

It wasn't until the turn of the twentieth century that another French mathematician, Jules Henri Poincaré, pointed out the limitations of Laplace's conjecture. Though Poincaré agreed with Laplace that a demon knowing absolutely everything about the present might determine everything about the past and future as well, he demonstrated that minor uncertainties in knowing the former would lead to vast errors in forecasting the latter. Because minor fluctuations exist in all measurements, Poincaré's findings severely weakened Laplace's hypothesis. As Poincaré wrote in 1903: "It may happen that small differences in the initial conditions produce very great ones in the final phenomena . . . prediction then becomes impossible."[7]

In the 1800s, advocates of determinism had not anticipated such limitations of knowledge. Even after Poincaré warned about the hidden power of small effects, few heeded his admonition—until chaos theory in the 1960s and 1970s proved him most prescient. For more than a century—between the age of Laplace and the advent of quantum mechanics and then chaotic dynamics—determinism's adherents steered the battering ram of classical Newtonian science, upholding Laplace's conviction that mechanistic reasoning could bash its way through any obstacle.

One might readily understand why determinism held sway for so long. As religious belief in direct divine influence over the world increasingly became anathema for many intellectuals, the times demanded a new faith. The solidity of determinism, claiming absolute knowledge and invariable order, seemed to occupy a niche that religion, for many, had ceased to fulfill. As faith in the power of equations continued to expand, the clockwork deity, ruler of the age of the machine, had become the new godhead to which rational, secular thinkers paid tribute.

The Beats of Different Determinisms

Determinism, historically, has by no means been a monolithic movement. Though Laplace's interpretation of Newtonian physics provided one of its shining moments, deterministic philosophy predates Laplace, and has survived in various forms until the present day. Movements as diverse as Calvinism, Darwinism, Marxism, Skinnerian behaviorism, and Freudian psychoanalysis each encompass a variation of this powerful system of beliefs.

Traditionally, determinism is defined in contrast with the more amorphous concept of free will. Essentially, the difference between the two beliefs resides in how much control individuals are thought to have over their own lives. For a strict determinist, the answer is none; destinies are guided by forces beyond one's reach. In the cosmic balance sheet, the sum of things past exactly preordains the totality of things to come—with no doctoring of the ledger (through conscious intervention) allowed.

In contrast, individuals advocating some form of free will believe that although the past shapes the present and the present guides the future, conscious minds might break this chain at any time. Processes such as artistic creativity, musical prowess and mathematical genius represent, for these advocates, emblems of spontaneous expression not beholden to any source (or, at least, not completely determined by prior events).

Could a powerful computer, fed with the entirety of information about Beethoven's genetic makeup, family upbringing, class background, hearing difficulties, and so on, have predicted his Fifth Symphony, note by note? Or did the resounding, anthem-like tones of that piece arise as a pure, impromptu expression of his art that no mechanism could have ever foreseen? Without being able to perform an objective study—re-creating the past, altering various factors, and comparing outcomes—one truly cannot solve such a riddle. Thus, one's answer purely depends on one's philosophy—determinists leaning one way and "free-willers" the other.

Many well-known thinkers have argued on either side of the free will versus determinism battle. John Dewey, one of America's most prominent political and educational democratic reformers, believed ardently that human behavior was solely determined by what he called "the push of the past." William James, on the other hand, a fellow member of the pragmatist movement, believed that nobody could fully anticipate the future, not even God. "That," he wrote," is what gives the palpitating reality to our moral life and makes it tingle . . . with so strange and elaborate an excitement."[8]

C. S. Lewis, the well-known Christian philosopher and author of beloved children's tales, thought that it was silly to proclaim "belief" in determinism, because the act of believing in something implies free choice. Therefore, he argued, determinists should not be smug in dismissing the process of free will.

The squabble between these two modes of thinking has precipitated significant religious rifts. Issues that arose at the time of the Reformation, when Catholic traditionalists faced Protestant dissenters, provide a prime example. Although both the Church and Reformers advocated forms of predestination, Reformers such as Martin Luther and especially John Calvin took much stricter positions about the matter than did members of the Church.

In traditional Catholic doctrine, one's fate after death—heaven or hell—depends on one's actions during one's lifetime, particularly on one's acceptance or rejection of Christ. Although good works, devout faith, and repentance for one's sins serve to open St. Peter's gates, wicked doings without contrition ensure that these doors remain locked. As a sentient being, created in God's image, one is blessed with the free choice of action. Each person elects to lead a life of good or evil, and enjoys or suffers the consequences of one's decision. Though God can foresee all, he withholds judgment until that choice is made. Exactly how God can always be correct in his forecasts, and yet allow complete free choice, has been a matter of considerable Church discussion for centuries. Nevertheless, the Church sees no fundamental contradiction between divine foreknowledge and free will.

In contrast, according to Calvinism and certain other Protestant creeds, humans have lost their free will in the act of original sin, and can regain it only through salvation. God, in his omniscience, foreordains who will be blessed and who will be damned. Because God's knowledge is perfect and immutable, human fates are sealed from the moment of birth. Hence, one cannot change one's destiny, but only have faith that one has been chosen. Most of the elect, according to such doctrines, eventually realize their distinction, and thankfully experience the fruit of God's special blessings. Only then do they inherit their birthright of willpower.

Some kinds of determinism—Laplace's interpretation and religious predestination, for example—support the maintenance of the status quo: an unchanging clockwork universe, a constant progression of souls into heaven and hell, and so forth. In these types, rhythmic mechanisms, be they material or divine, ensure that the order of the cosmos remains perpetually the same. Other deterministic varieties, in comparison, purport that change is inevitable. Automatic processes, scientific or otherwise, ensure that time will march forward to the beat of continuous development. Such evolutionary movements include Marxism and Darwinism.

Some political scientists have referred to Marxism as a kind of religion. Based on its complex history, one can understand why. From its inception in the nineteenth century, the movement founded by Karl Marx and Friedrich Engels has supported a bevy of zealous believers, iconic figures, and cultlike splinter groups clashing over doctrines of faith. And Marxism has its paradise as well, its prophesized future period, known as communism, when class conflicts will cease and history as we know it will end.

The driving force behind social change, according to Marx and Engels, is an evolutionary process called economic determinism. According to this standpoint, history might be characterized as a struggle between economic classes, in which oppressed groups fight with dominant groups for control, until the former manage to overthrow the latter. In time, those once trampled upon assume the role of their former masters, donning the lofty boots of power and stamping down the ambitions of a new underclass. For example, the merchant class, derided by feudal lords during the Middle Ages, has, since the start of the industrial age, assumed leadership of society and exerted economic authority over workers.

This developmental process continues, according to Marxist theory, until the workers themselves assume control over the means of production (mills, factories, energy works, and so forth). Because no other ruling class might supplant the workers, the machinery of history grinds to a finale.

With neither mention of free will nor of individual influence, it's clear why philosophers characterize Marxism as a deterministic approach. Global economic necessity, not personal free choice, ignites its spark for change. A lone individual has as little chance of changing history as a briny bubble has for preventing the sea from crashing over an ineffective dam.

For example, according to many Marxists, Roosevelt should receive no personal credit for advancing the New Deal. Rather than respond to an air of crisis, and compassionately develop new programs for the unemployed, he acted merely as a representative for the ruling class to preserve the status quo at all costs. From the Marxist perspective, history should not judge individuals for their so-called choices; rather, it should examine the economic forces that compelled them to make these decisions.

Marxism, and other deterministic sociological and economic movements, claim to have discovered the scientific mechanism for historical change. Because scientific investigation typically involves experimentation, however, and societies, in their entirety, cannot readily be dissected like fetal pigs nor tampered with like genetic material, objective analysis of these movements' assertions often is impossible. Stated another way, because of science's freer hand in launching controlled experiments in biology—compared to sociology, for instance—science's understanding of biological order is far superior to its knowledge of social order. There is no comprehensive theory in politics that provides as elegant and simple an explanation as Darwin's does for the development and diversity of life. Nor can any sociological approach muster the predictive power of Darwinism in modeling the growth of complexity.

Nature Versus Nurture

Walking through a tropical rain forest, one cannot help be dazzled by its rich potpourri of sounds, scents, and colors—the shrill cry of a macaw, the fragrance of exotic plumeria, the rainbow hues of orchids. With millions of species all around, nature shines in beautiful redundancy.

Though such remarkable diversity satisfies the senses, no doubt, one might wonder what constitutes its utilitarian purpose. Why are there so many varieties of orchids, for example, and not just one or two? Is their wide range of attractive colors designed just to please the eye, or does it stem from more pragmatic reasons? Or, conversely, might it just be due to chance? Furthermore, why do some species seem to thrive in certain regions, while others tend to die out?

In 1859, English naturalist Charles Darwin brilliantly addressed these questions in his groundbreaking treatise, *On the Origin of the Species by Means of Natural Selection.* Starting on humble ground, by speculating about variation among breeds of pigeons, Darwin rose to the task of explaining how complex species evolve from simpler organisms though differentiation and competition. In doing so, he furnished, for living beings, the same kind of simple behavioral mechanism that Kepler and Newton provided for celestial bodies.

The first step of the Darwinian process involves variation of characteristics from one generation to the next—for example, a white dog giving birth to one with black spots. Darwin did not explain how these changes come about; today we know these arise through mutation and crossover of genetic material.

The next stage involves competition. Within a given environment, a wide range of creatures fight for sustenance and shelter, attempting to live long enough to produce healthy offspring. The more babies produced and the more viable they are, the more successful a particular organism's life can be said to be.

Some of the variations described in the first step serve to maximize a being's chances of success. For instance, a color change providing superior camouflage might enhance the odds of an animal's survival. Other alterations, however, might prove detrimental. To consider an extreme example, a creature born without reproductive organs would hardly be blessed with an advantage for producing young.

Evolution occurs because favorable variations tend to flourish, and unfavorable ones tend to die out over time. Ill-adapted species often become extinct—especially when their environment changes for the worse. Those beings that do survive through the eons tend to be the hardiest, or else they are lucky enough—like the duck-billed platypus—to have few competitors.

Darwin's model, in its basic form, constitutes a looser kind of determinism than, say, the perpetual motion machinery of Laplace. An individual organism, particularly if it is complex enough, still possesses a measure of free will. As long as it follows its instincts to protect itself and its young, it might live its life as it pleases.

Stricter variations of Darwinism, however, assert that virtually every sort of animal behavior—even human conduct—is biologically preprogrammed. In theories of genetic determinism, people inherit all measure of characteristics from their ancestors—their strengths and weaknesses, their emotional dispositions, even their political and interpersonal behaviors. Hypothetically, according to these views, future geneticists will be able to provide expectant mothers with accurate profiles of their fetus's predicted physical, social, and psychological attributes.

Many behavioral scientists, however, take a radically different tack. In the famous "nature versus nurture" dilemma, they favor the latter as a defining cause, and argue that personality is largely shaped by early childhood experiences. B. F. Skinner, founder of the movement known as behaviorism, conducted numerous conditioning experiments indicating how subtle rewards and punishments influence the development of children, molding their embryonic dispositions. A smile from one's parents, for instance, might induce a child to pick up a book, and stimulate a lifetime of reading pleasure.

Today, virtually no psychologists would consider themselves pure Skinnerians. Almost all would argue that personality is shaped by both genetic and acquired attributes. The main question that divides various movements—sparring with each other over wine and cheese—concerns where to draw the line. How much of the mind is nature, and how much is nurture, continues to be an issue of hot contention.

Freudian Slips

It is a tribute to the power of Newtonian and Darwinian notions that, at the turn of the twentieth century, many scientists believed they had little left to do except tidy up loose ends. All physical and living systems appeared to operate on predictable mechanistic principles. It was only a matter of time before everything under the Sun—and beyond the Sun as well—could be wholly understood down to its finest detail.

Randomness, both in science and in society, seemed to be a vestige of far more brutal times. Order and rigidity became high aspirations, from scientific proofs to social etiquette. Britain, for instance, marched to the regimented beat of its Victorian era, a time when modesty and formality reached its zenith. Only in dreams and fantasies might austere Victorian socialites experience flights of pure nonsense. That's why Lewis Carroll—as an eccentric and an aberra-

FIGURE 2.8 Sigmund Freud (1856–
1939), father of modern psychology and
founder of the psychoanalytic method (courtesy
of the Library of Congress).

tion—was considered so amusing. In writings such as *Alice in Wonderland,* he mocked rigid custom and brought momentary respectability to the absurdity of dreamland.

Considering the curtain of modesty drawn between the constraints of reality and the freedom of fantasy, imagine the shrieks of the ladies and gents when Viennese psychologist Sigmund Freud lifted up this protective divider and let the light through. In his masterful *The Interpretation of Dreams,* published in 1900, he argues that repressed desires manifest themselves as distorted images in dreams. Jokes, nervous tics, off-hand remarks, and even slips of the tongue often reveal hidden wishes too dark to express directly.

According to the principle of psychic determinism, a tenet of strict Freudians, everything one says or does reflects either conscious or unconscious thoughts and desires. If a speaker forgets the name of someone she is about to introduce, invariably there is good reason for her absent-mindedness. Perhaps his profile reminds her of her father, for whom she maintains ambivalent feelings. By forgetting the look-alike's name, she has prevented her father's memory from coming to the fore.

Charles Brenner, in his *Elementary Textbook of Psychoanalysis,* the classic handbook for a generation of psychotherapists, captures well the deterministic view of psychology:

> In the mind, as in physical nature around us, nothing happens by chance, or in a random way. Each psychic event is determined by the ones that preceded it. Events in our mental lives that may seem to be random and unrelated to what went on before are only apparently so. In fact mental phenomena are no more capable of such a lack of causal connection to the ones which preceded them than are physical ones.[9]

Freud's theories scraped the veneer of innocence from the wooden occupants of richly furnished Victorian sitting rooms. Lads or lasses who, in principle, shared Anthony Comstock's ideal of absolute discretion, were surely mortified to learn that their jokes and misstatements revealed to the world their secret, libidinous selves. In an emphasis on the symbolism of dreams unmatched since the days of Delphi, nervous souls sifted through their own nocturnal imagery, worried about repressed desires threatening to escape. Robert Louis Stevenson's *Dr. Jekyll and Mr. Hyde,* written in 1886, eerily anticipated such monstrous fears.

Paradoxically, though Freud's work attempted to bring a Newtonian, mechanistic slant to the inner workings of the mind, it focused sharp attention on the decidedly messy subtext of human thought. If, as Brenner writes, each thought is inextricably linked with all others, this complex web of related experiences seems as remote from the simple, clockwork image of the planets as the global telephone network does from a string and two cups. Rather than represent a narrow, focused hallway leading—as Newtonian dynamics was thought to do—to ultimate predictability and the end of science, Freudian psychology constituted an expansive highway into new uncharted vistas of labyrinthine complexity.

Five years after Freud published his bold attempt to decipher the patterns of dreams and thereby map the hidden mechanisms of the mind, Einstein introduced his equally ambitious scheme to unravel the behavior of near–light speed objects and thus fathom the structure of space and time itself. Later he was to place his theory of special relativity within the context of a broader approach called general relativity. Einstein's models came to supercede Newtonian physics, particularly for situations involving high speeds and strong gravitation.

As in the case of Freudian psychology, Einsteinian relativity aspires to provide deterministic laws resulting in absolute predictability. Indeed, Einstein's equations are fully deterministic, providing indispensable tools for examining the past, present, and future of the cosmos. Yet, as interpreters of Einstein's work have come to discover, the structure of the universe appears to harbor twisted, paradoxical elements that challenge explanation. Like Freud's theories of the mind, relativity theory has spurred on an increasing recognition that the world is far more complex than previously thought and that many of its mechanisms are unseen and even beyond perception. The mysterious bodies called black holes, and similar highly compact objects described by relativistic physics, offer examples in which time itself seems to curve, connect up with itself, even dead end. Prediction typically assumes that cause precedes effect—not necessarily the case in certain unusual regions of the cosmos. Reconciling the coiled-up nature of time under extreme relativistic circumstances with the conventional progression of causality remains an elusive goal of modern physics.

LIGHT FLIGHTS

The Einsteinian Future

*Now, here, you see, it takes all the running
you can do, to keep in the same place. If you
want to get somewhere else, you must run at
least twice as fast as that!*

—Lewis Carroll
Through the Looking-Glass

The Fourth Dimension

Looking out the window of an airplane, shortly after takeoff, the world
below seems to flatten, enlarge, and transform itself into something resembling a map. Communities appear to collapse into simple geometric forms: houses into squares, fields into rectangles and trapezoids,
roads into crisscrossing lines. A moment's glance might capture a
whole municipality, reduced by mile-high vantage into a panoramic
snapshot.

Imagine being able to escape from reality itself, lifting off from the
totality of space and time. Picture soaring above everything in existence—past, present, and future—and instantly observing the history
of the universe spread out like a road map. No feature of creation
would be hidden from one's view.

Impossible? Certainly for us mortals—under ordinary circumstances at least. Only through our imaginations could we attempt this feat. We could try to picture all of space at a particular moment as a box (perhaps infinite), and all of space at the next moment as another box, and so forth. Then, by mentally stacking these cartons one after another in temporal order, we could attempt to envision reality for all times. The main conceptual difficulty would be to choose the direction of stacking. Because the boxes occupy the three spatial dimensions, we would have to stack them in a fourth dimension, perpendicular to all three. The fourth dimension would thus represent time.

In the late nineteenth century, a few of the more imaginative members of the scientific and mathematical communities speculated that time was a dimension every bit as real as length, width, and height. Lacking basis in firm evidence, they built their conjectures largely on democratic sentiments: Why not grant time an equal place at the table along with its three spatial brethren?

These speculations inspired H. G. Wells, who incorporated them into the scientific pretext of his classic novella, *The Time Machine*, published in 1895. As he wrote in the opening pages: "There is no difference between time and any of the three dimensions of space except that our consciousness moves along it."[1]

Placing time on an equal footing with space raises intriguing issues about the meaning of destiny. Does the future already exist, inaccessible to us, but potentially observable under special circumstances? Could an imaginary being, situated outside the confines of space and time as we know them to be, theoretically view all of history at once?

The possibility that all of space and time forms an invariant whole—an idea known as the "block universe" approach—seems to imply a stronger form of determinism than even Laplace suggested. Although in Laplace's theory complete knowledge of the universe at any given moment would enable one to forecast the future perfectly, one could never verify such a prediction by actually experiencing the times to come. Physically, as well as experientially, history would unroll gradually, revealing its contents in dribbles and drabs.

The "block universe" notion, on the other hand, suggests that the cosmic chronicle simply exists—a complete, never-changing entity. Each moment is related to the next like adjacent frames of a reel of film. The flow of time represents a mere illusion—a feature of the way humans (or living beings in general) view the world.

Taken to an extreme, this philosophy implies that we are all acting in movies with screenplays written long ago. Our mental "Tele PrompTers" feed us the words that we will say, and we inevitably recite them without alteration. Our actions and emotions, as well as the cast of characters that we will encounter at each moment of our lives, are similarly penned into our scripts. Life is interesting and novel only because we lack the capacity to foresee how our dramas will turn out. Moreover, we enjoy (or don't enjoy) our existence, because our feelings have been already been scripted.

Imagine how dull and frustrating it would be if we somehow gained access to the entire cosmic play, knew all our lines in advance, but were powerless to do anything about them. We'd be like Cassandras, condemned to foresee all of our misfortunes, and then be forced to experience them without possibility of change. What would be the point of prediction if the future were already written?

Some of the most intriguing depictions of this theme—trying to maintain sanity and hope in light of preordained outcomes—have been written by the renowned Argentine author of poetry and prose, Jorge Luis Borges. In his enigmatic story, "The Aleph," Borges describes a mysterious orb through which "all the places in the world are found, seen from every angle."[2] Though it is a small sphere, about an inch in diameter, the vast expanse of space and time might be seen within it. After the narrator of the tale peers into the crystal ball, he finds it difficult, for a time, to find pleasure in ordinary living. All faces and things look familiar to him, and he fears that life will hold no more surprises. He loses many nights of sleep mulling over the kaleidoscope of images he has seen. Happily, though, amnesia eventually allows him welcome relief. Omniscience, it seems, has been too heavy a burden for his fragile soul to bear, and forgetfulness has provided the only cure.

This Borgesian motif has found recent expression in the movie *Pi*, where a young mathematician drives himself to the brink while contemplating numerical patterns found in both the Bible and in the stock market. He becomes obsessed with finding a sequence of digits that he believes forms the key to understanding the purpose and fate of the universe. His mad quest draws the interest of Kabbalists and investment brokers alike, who pursue him and add to his paranoia. Once he thinks he's unraveled the secret code of the cosmos, his sanity crumbles altogether. Knowledge of universal destiny proves too great a load for his weary brain. Only by putting such thoughts aside, the film suggests, does he finally learn how to relax and begin to enjoy life.

Fortunately for our mental health, the ability to perceive the entire "block universe" remains far beyond the scope of mortal possibility. Only a god might, as Shakespeare put it "count [him]self a King of infinite space." For now, our limited minds can only predict the likelihood of events to come—never knowing the future with certainty.

No individual contributed more to our current understanding of space and time than Albert Einstein. In his theory of special relativity, Einstein not only showed that space and time form a unity—dubbed "spacetime" by the Russian mathematician Hermann Minkowski—he also precisely delineated the limits of communication within this medium. The speed of light, Einstein demonstrated, sets the pace of causal interaction. One event could influence another only if the distance between them is close enough that light could travel from the first to the second in the intervening time. For example, a sudden release of energy on the Sun could affect the weather on Earth the next day—or even the next hour—because light would have sufficient time to bridge the gap between the two bodies. However, the solar flare could not cause a terrestrial effect just one second later; causal influence, traveling at light speed, could not proceed fast enough.

Within the arena of spacetime, the boundary between events within causal range of a given incident and those out of range of that incident form a four-dimensional geometric structure analogous to a cone. This "light cone" traces all possible paths light can make through spacetime, starting or ending at the occurrence under consideration. For example, the light cone for a solar flare comprises the points in space where the radiation from that flare could reach at any given time. The flare could affect only locations on or within the set's boundary (reachable in the speed of light or less); other locales would be out of range. As time goes on, these regions of influence widen; hence the conelike structure.

Even if we live in a "block universe," we cannot step out of spacetime and view it as a whole. To predict the future, we are limited to the information our senses and instruments can gather. Einstein's theory tells us that reality is a labyrinth of overlapping zones of influence.

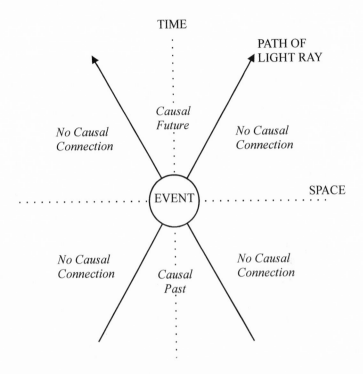

FIGURE 3.1 *This light-cone diagram depicts the possible types of causal relationships between an event and other regions of spacetime. Only one of the three spatial axes is shown. A true rendition would be four-dimensional.*

These delineate exactly what can be known about the past and, as a consequence, what can be ascertained about the times ahead.

If these zones of influence were regular and static, then prediction in relativity would be fairly straightforward. We would always recognize the extent of what can be known. As special relativity informs us, however, observers traveling at different speeds possess distinct perspectives on space and time. General relativity, its successor theory, complicates the situation even more, showing that light cones bend in the presence of mass. As Einstein demonstrated, the material distribution of the universe alters its topology, changes the path of its light rays, and modifies its flow of information. Consequently, Einsteinian spacetime is an astonishingly intricate array of causal influences—making it hard to know the limits of forecasting. Let's examine what motivated Einstein to develop such an elaborate approach.

Flexible Moments

Science strives for the simplest possible description of any natural phenomenon. Generally, only when contradictions arise do complex models supercede more basic theories. In 1905, Einstein introduced special relativity as a means of resolving a troubling inconsistency in physics. He wished to reconcile the constancy of the speed of light in space with the relative behavior of other types of motion. No matter how fast or slow one moves, he wondered, why does light always seem to pass by at the same rate?

The notion that the speed of light is constant in a vacuum dates back to the foundations of electromagnetic theory. In the nineteenth century, physicist James Clerk Maxwell theorized that the speed of light depended only on the intrinsic properties of electric and magnetic fields. Light, his unified theory suggested, did not need a substance through which to propagate; rather, it traveled through space at a fixed rate owing to its own internal dynamics. Still, many scientists refused to believe that something could move through nothing, and proposed a "luminiferous aether" through which light waves vibrated.

To test the properties of this aether, experimentalists Albert Michelson and Edward Morley rigged up a clever contraption designed to test the velocity of a beam of light. Using a series of mirrors they split the beam in two, and forced each to follow separate, but equidistant, paths. Then they reunited the segments, and—postulating that the aether must have a particular direction—determined whether the light took longer to travel with the aether or against the aether.

Their experiment was analogous to measuring ocean currents in the Atlantic by gauging the speed of a ship as it voyages the same path twice, first eastward and then westward. All things being equal, as it journeys with the current it should sail faster than when it goes against the current. By comparing the times taken each way, and using this information to compute the difference in the ship's speed, one might estimate the velocity of the current. In a similar manner, the researchers hoped to single out the effects of the aether.

FIGURE 3.2 Albert Einstein (1879–1955). Einstein's special and general theories of relativity radically transformed the way scientists envision the future (courtesy of the AIP Emilio Segre Visual Archives).

To Michelson and Morley's astonishment, they recorded absolutely no difference in light's speed no matter which way it traveled relative to the aether. They concluded that no such substance exists, and that light's speed in empty space never changes. Unlike ocean waves or sound waves, light waves, they found, can travel without a medium.

Einstein was perplexed by the seeming contradiction between light's invariant motion and the relative behavior of other objects in classical mechanics. If the speed of light in a vacuum is always the same, the velocity of the person taking the measurement must not matter. In Newtonian physics, however, all speeds must be considered relative to the motion of the observer. The following example illustrates this principle: Imagine someone standing on the side of a highway and watching a truck and a car zoom by on two different lanes. Traveling next to each other at the same speed on two different lanes on a highway, it would seem to the truck driver that the car isn't moving at all. However, someone standing on the side of the road would see both the car and the truck zoom by. Therefore, the speed of the car would seem different to the fixed and moving observers. This idea, that the speed of an object is observer-dependent, is called Galilean relativity and forms an essential feature of classical physics.

Now consider what would happen if the truck driver, instead of traveling alongside a car, tried to do the same with a beam of light. Let's say he is driving a fusion-driven, high-speed space truck, to make his goal more realistic. According to Galilean relativity, as he approached his target velocity—light-speed—he would start to perceive that the light isn't moving (or at least that it's moving very slowly). But, according to Michelson and Morley's results, all observers should measure light to be moving at the same speed. Light should never appear to be standing still. In short, a contradiction would arise between two different methods of gauging light's relative speed—one approach finding it to be zero; the other, to be the same as usual.

In his theory of special relativity, Einstein brilliantly resolved this dilemma. He replaced the notions of absolute time and absolute space, central to the empirical foundations of Newtonian mechanics, with the concepts of observer-dependent time and observer-dependent space. That is, instead of distance and duration being fixed quantities independent of the speed of the measurer, they become relative quantities that take on different values depending on the measurer's speed.

Let us consider how Einstein's theory handles the question of time for an object moving at a constant speed relative to a nonmoving observer. To simplify matters, let's call the object's point of reference the "comoving" frame, and the static vantage point the "fixed" frame. In the process known as time dilation, special relativity mandates that, according to the fixed observer, a comoving clock (one attached to the moving object) would tick more slowly than a fixed clock (a watch on the fixed observer's wrist, for example). That is, time seems to slow down for the comoving frame, relative to the fixed frame.

As incredible as the time dilation effect sounds, it is quite real, and altogether testable. In fact, extremely precise atomic clocks positioned on supersonic aircraft have noted minute, but significant, differences from the time on the ground. To capture larger discrepancies, one must perform such experiments on much faster forms of flight. Only ultra-high-speed space travel, far faster than any present-day spacecraft have ever flown, would generate the speeds necessary for a significant effect—measurable in months or years, not just fractions of a second. (This is on a human scale. On a subatomic scale, particles moving near light speeds have significant, testable relativistic discrepancies in decay rates.) The reason for this requirement is that large-scale time dilation only occurs when an object travels close to the speed of light.

Imagine an astronaut leaving Earth at an enormous speed—a sizable fraction of the velocity of light. He journeys into space and returns to Earth in one year, according to his ship's clock. Rather than exploring new worlds or scoping out new astronomical features, his main mission is to examine time dilation effects. Testing special relativity, he wishes to compare his own temporal experience with that of those he left behind.

At 90 percent of the speed of light, for example, time would advance for him more than twice as slowly as it would on Earth. When he lands he would probably notice few changes, save the fact that the calendar has skipped ahead an extra fifteen months or so. Because presumably his family and friends would look and act pretty much the same, and the world would have changed very little, he would almost certainly be able to resume the life he led before his journey. Relativistic effects scarcely would have affected him.

Traveling at 99 percent of light speed, the difference would be more noticeable. The astronaut would age only about a single year, while his friends back on Earth would pass through seven whole years of life. Finally, at 99.99 percent of the speed of light the ratio between Earth years and spaceship years would be a whopping seventy times slower. When he steps through the front door of his house, he would be more likely to encounter his grown grandchildren or great-grandchildren than to reunite with any of his contemporaries.

It is straightforward to see how the time dilation effect would resolve the truck driver example discussed earlier. Once again, let us imagine him trying to ride alongside a beam of light in his space truck, only this time, we'll take relativity into account. As he moves faster and faster, the atomic clock located in the front panel of his vehicle would start to slow down, compared to what it would read if he weren't moving. Because everything else in the cabin would slow down in time as well, including his thoughts and his metabolism, he would not notice the difference.

What if he now looked through his windshield (a *solar* windshield, that is) at the luminous beam. He would see a stream of photons zooming by at the standard speed of light—exactly as he would have if he were parked. Although his truck would be moving extremely fast, his clock would be running correspondingly slow, relative to its terrestrial rate. The two effects (high speed and time dilation) would cancel, and the light particles would appear to be keeping their ordinary pace.

Would the truck driver ever be able to catch up to the light stream? The answer is negative—according to the driver, as well as to all other observers. According to what we've just stated, it's clear why the driver would observe that he could never outpace the beam; his laggard clock would not let him. To explain why this would also be true from everyone else's point of view, one must look at the velocity limitations posed by Einstein's theory.

Tachyonic Antitelephones

One of the most important consequences of special relativity is a prohibition against subluminal objects reaching or exceeding light speed. Einstein showed that as a physical body would approach the velocity of light, it would take more and more energy for it to accelerate. To reach light speed it would take an infinite quantity of energy. Because such power would manifestly be unobtainable, reaching light speed from below would be impossible.

Objects already at light speed would not be subject to this prohibition. An X-ray, for instance, naturally moving at the speed of light, would require no additional energy to maintain the same velocity. In short, once subluminal, always subluminal; once at light speed, always at light speed.

What about travel at supraluminal speeds: velocities faster than light? Although Einstein did not anticipate such a possibility, his theory does not explicitly rule it out. As long as an object does not accelerate *up* to light speed, special relativity does not forbid it from going *over* light's speed. Strangely, though the cosmic police post strict laws against speeding, those already breaking the speed limit seem to be exempt.

The term *tachyon,* referring to hypothetical supraluminal particles, was coined by Columbia physicist Gerald Feinberg in the early 1960s. He demonstrated how such objects might exist in nature, wholly undetected, engaged in the mirror image of ordinary behavior.

Much like ordinary bodies cannot ascend to the level of light speed, tachyons would not be able to descend to that platform. In other words, once a tachyon, always a tachyon.

If tachyons existed in nature, they would have most unusual properties. In contrast to known natural objects, their masses would be imaginary; that is, multiples of the square root of negative one. Adding energy to one of these creatures would cause it to move slower and slower (but still faster than light). Like an unwanted lodger, the more calories they would consume, the more lethargic they would become (though they still would move fast enough to dash to the refrigerator unseen).

Furthermore, tachyons would exhibit the bizarre behavior of always traveling backward in time. With their high speeds, they would beat out light signals emitted by physical occurrences, and thereby sense events in reverse order. A tachyon would, for example, come into contact with the blast of an explosion before it encountered the light from the spark that triggered the blast.

In an investigation of these ideas, published in an article entitled "The Tachyonic Antitelephone," physicists G. A. Benford, D. L. Book, and W. A. Newcomb demonstrate how tachyons would create knotted contradictions, impossible to resolve. They imagine using streams of these speedy particles to engage in backward-in-time communication. "Tachyonic Antitelephones" would enable one to chat with people in the past—assuming they possessed appropriate receivers. These devices would also allow those of the present to receive "premonitions" of the future by means of tachyonic messages.

Such "telephones" would be useful; a bored woman waiting for a delayed bus could call and "remind" an earlier version of herself, before she leaves the house, to bring along a copy of *War and Peace.* Through the wonder of tachyonic transmission, she would then have something to read.

But if she has reading material already, why would she make the call? If she doesn't make the call, though, she would not obtain the book. Hence, reality would contain two contradictory possibilities: either she has the book and does not phone herself, or she doesn't have the book and phones herself.

Benford, Book, and Newcomb, after citing similar examples, conclude that experiments searching for tachyons are doomed to fail. Indeed, in the decades since tachyons were first proposed, none have been found.

The Theater of Spacetime

Postulating the principle of time dilation—and of a related phenomenon, known as length contraction, in which high-speed objects appear to shrink in the direction of their motion—Einstein unveiled a profoundly original perspective on time and space. Breaking the cadence of the Newtonian mantra "absolute space and absolute time," chanted by scientists for centuries, he showed that these parameters no longer could be said to form independent gauges of reality. Rather, one must consider spacetime as a *gestalt,* an inseparable four-dimensional entity. Distances and durations no longer mattered in and of themselves; what counted was the four-dimensional, spacetime separation between events.

In 1908, Hermann Minkowski formalized this concept. In an influential lecture, he spoke of space and time as mere "shadows" of the more fundamental concept of spacetime. He demonstrated that the laws of physics, including the equations of electromagnetism and mechanics, could be elegantly rewritten in terms of four-dimensional spacetime coordinates, and proved these revised expressions fully equivalent to Einsteinian special relativity. In essence, he constructed a solid mathematical foundation for the notion of time as the fourth dimension.

Einstein viewed Minkowski's approach as an opportunity to solve a long-standing problem in physics: the issue of describing gravitational interactions. Finding inadequate the Newtonian notion that gravity is a force that acts between two objects over a distance, Einstein sought to replace that idea with a local representation; namely, a description that involves actions taking place at single points. Seeing how Maxwell's equations provided a brilliant local explanation of electromagnetism—charges and currents respond to "fields" created by other charges and currents—Einstein wanted to repeat the same feat in characterizing gravitation.

In 1916, Einstein introduced his theory of general relativity, a way of describing gravity by means of spacetime distortions. He viewed the geometry of the universe as an active dynamical element, reflecting the influence of mass in a particular region. Depending upon the distribution of matter, reality could be stretched and shaped, twisted, and even torn. In advancing these ideas, he broke sharply with classically rigid notions of space and time.

According to the Newtonian perspective, space and time form a kind of traditional boxed-in theater in which the drama of the universe is performed. Assorted players make their entrances and exits, engaging in myriad interactions, but the rigid cosmic backdrop, painted in indelible ink, remains unaltered. It is as if the actors are forbidden by custom and contract to fiddle around with the set; nothing might be changed in any way.

Einstein's general relativistic proposal, on the other hand, resembles a more contemporary drama in which the players continually alter elements of their stage. From performance to performance, not only do the actors and their relationships differ, but the background and props change as well. The transformations in scenery profoundly influence the movements of the performers. These actors, in turn, affect new alterations in the setting around them. Thus, a symbiotic interplay between cast and stage propels the drama ever forward.

Einstein's equations of general relativity are striking for their succinctness, and remarkable for their maddening range of solutions. They relate two mathematical entities, called tensors, that possess particular kinds of transformational properties.

On the left side of his equations stands the Einstein tensor, a function of something called the metric. The metric is a shorthand description of the spacetime distances between all pairs of points in the universe. Like the topographic maps handed out at national parks, it characterizes how far one has to scurry to travel from one point to another. For cases in which spacetime is perfectly flat, then this description is easy: like cities on the Great Plains, each set of points can readily be connected by a perfectly straight line. But try walking a perfectly straight line across the Andes or the Himalayas, and one can see why a more complicated map of spacetime is required for curved geometries.

On the Einstein equations' right-hand side lies a second mathematical object, the stress-energy tensor. This entity serves to summarize the quantities and properties of matter and energy in a particular region. For example, a section of spacetime might contain a dense star, a tiny planet, or simply a dilute pool of radiation.

By equating these two tensors, Einstein quantified how the presence of mass impacts the structure of spacetime. Heavy materials, according to these relationships, serve to distort the shortest paths between spacetime points, much in the same way that avalanches, rock slides, and earthquakes alter routes through the mountains. Straight lines bend, and former shortcuts become no longer as short.

In the absence of matter—these equations show—spacetime is smooth. Once uneven material distributions are taken into account, spacetime becomes bumpier; the more mass in a given region, the more distortion. The curvature of spacetime, in turn, changes its metric—altering the closest (straight-line) connections between each pair of points. Photons, and all other particles, must consequently follow modified routes through the cosmos.

General relativity explains why planets orbit their mother stars. Like a boulder on soft soil, each star, with its bulk, creates a pit of spacetime around it. Within these recessed areas, the planets surrounding that star remain trapped, possessing only the freedom to revolve around it.

Even the paths of light rays bend in the presence of a massive body. If the Sun lies between a star and Earth, for example, the starlight curves before we see it. The Sun's presence distorts the spacetime region around it, curving the hitherto straight trajectory of the light.

In 1919, this remarkable result was confirmed by two teams of British astronomers led by relativity expert Arthur Eddington. On expeditions to the Southern Hemisphere, they measured the angular position of stars situated close to the Sun's rim. The researchers took these readings during a solar eclipse, so that sunlight would not blur the view. Remarkably, they found a small but significant difference in the stellar bodies' locations, compared to times in which the Sun was not between the stars and Earth. The bending of the starlight closely matched the amount anticipated by general relativity. With this discovery, Einstein's theory won overwhelming acceptance in the scientific community as the standard model of gravitation.

Around the same time, researchers began to map the terrain of solutions to Einstein's equations. Like the first explorers encountering the Grand Canyon, they were astonished by the intricacy of the topographies they found. Spacetime, they discovered, could mold itself into myriad possible shapes—some even defying physical intuition about the law of cause and effect.

Predicting the future, once seen as an exercise in picturing the inevitable consequences of an absolute present, no longer could be viewed in such simple terms. The malleability of spacetime meant that the concepts of past, present, and future lost their universal quality. Like a slab of meat, reality could be "carved" in many different ways, leading to distinct representations of time. No single denizen of the cosmos could claim that his perspective on what is "now," what is "later," and how much time passed in between was the only correct way of looking at things; others could rightly argue for completely different views.

In short, theorists have come to realize that although general relativity provides a fully deterministic description of how a particular model of spacetime might "evolve" (its spatial part changing over time), it allows for such a wide range of solutions that prediction becomes a complex matter. Even when physical constraints are fairly specific, mathematical results are often ambiguous. Moreover, in many cases, researchers offer conflicting interpretations of the maze of models found. Only by making broad assumptions (that spacetime in our vicinity is fairly typical, for example) have scholars working in general relativity been able to rule out large categories of solutions and make progress in the field.

Into the Vortex of Darkness

Perhaps no other natural phenomenon reveals more about the jumbling effect that general relativity has upon linear notions of destiny than the gravitationally powerful, ultradense relics of supermassive stars: the mysterious objects known as black holes. Within the confines of such bodies, time literally turns on its side, and history's scope reduces to a moment's glance. Before we discuss the bizarre properties of these craters in the cosmic terrain, let's examine how they are created.

Black holes represent the final step in a stellar death march that lasts millions of years. When, after a lifetime of billions of years, a star three times or greater the mass of the Sun reaches old age, its central core starts to run out of fuel and begins to shrink. Having largely depleted its primary source of power, the fusion of hydrogen and other light elements into heavier atoms, the core contracts to exploit its own gravitational energy. Meanwhile, the outer envelope of the star, no longer held tightly by the core, starts to swell. The star grows bigger and bigger, becoming an object known as a supergiant.

After a certain point, the core's remaining fuel reserves run out. No longer able to support itself, it commences a sudden and explosive contraction—blasting most of the star out into space in a fiery supernova burst. The shrunken center—all that remains of the star—draws in tightly to itself, crushing its material into a thick pulp. Strong repulsive forces kick in. These represent the tendency of the neutrons and protons within the core to resist being pressed together. For some objects, called neutron stars, these forces wear out gravity and block the contraction from proceeding any further. Neutron stars are highly dense bodies composed of extremely tightly packed neutrons. Lacking enough mass to complete their collapse, they are, in a sense, failed black holes.

If, on the other hand, the gravitational forces are great enough to overcome all resistance, then the contraction continues until the very particles that form the core are completely destroyed. Even the sturdy neutrons are smashed to smithereens. The star is dead; long live the black hole!

As with mortal man, upon death, stars gain a shroud of anonymity. Stellar features become cloaked, until little of their original identities are left. By the time the former shining glory has become a lightless and lifeless carcass, only very general properties such as mass, spin, and net charge remain identifiable features. That is why, according to a well-known theorem, black holes are said to have "no hair." Like expressionless faces crowned with shorn pates, they are hard to tell apart.

Black holes present blank visages to the world. They cannot be seen directly because of the effect upon light of their troubling inner structures. Their concentrated ballast bursts through the planking of spacetime, creating large rifts in its flooring. When light (or anything else) enters such a gap, it inevitably follows the natural path downward— with no chance of escape. All previously straight lines in the vicinity now bend into the crag, and along one of those curving paths is where the light must invariably go. Thus no radiance emanates from the dead star's underlayer (save renegade particles that manage to escape through quantum tunneling, a process we'll discuss).

Each black hole possesses an invisible boundary designating its "point of no return." Called the event horizon, this layer demarcates the region where nothing that enters can escape. Put another way, within this zone escape velocities exceed the speed of light—physical impossibilities. If an object passes fairly close to a black hole, but does not breach this boundary, by traveling fast enough it can still evade the dark object's grip. If it has already passed through the event horizon, however, like Milton's traitorous angel, it is doomed to fall downward into the bottomless pit.

A black hole's event horizon represents the very edge of predictability. Concealing from the outside world all knowledge of its interior, it reminds us that scientific prediction sometimes faces impregnable barriers. Only those unlucky enough to venture inside the relic star could learn the secrets protected within its fortress. Remaining safely outside, one could only speculate what such a ghastly voyage would be like.

Woe to the space traveler who enters the bleak portal of a black hole. From the moment he has passed within the confines of its event horizon, he would become inextricably drawn toward its dead center—a point of infinite density known as a singularity. Crushed into anonymity, his very atoms would exude into sheer oblivion—vanishing into a nameless place in which time and space simply come to a dead end.

For the majority of black holes—those produced by simple stellar collapse—a hapless intruder would have little time to think about his predicament. Tidal forces would rip him apart long before he collided with the singularity's infinitely thick wall—in many cases, even before he passed through the event horizon. The unequal actions of the dark object's gravity would elongate him like a rubber band in the direction that he is falling, and crush him in all other directions. With such enormous strain on his body and craft, he would be pulverized in a flash.

An average-size black hole (around ten solar masses, say) would hardly seem like an ideal place to test the precepts of general relativity. While being stretched like spaghetti, there would be scarcely enough time to scream, let alone to contemplate the enigmas of Einstein's theories.

Fortunately, for the purposes of scientific inquiry, much larger black holes do exist—ones that would grant explorers more time to think about their experiences, observing all the marvelous relativistic effects, before they are crushed. For example, in 1994 astronomers using the Hubble Space Telescope observed a supermassive black hole, estimated to be 2 to 3 billion times the mass of the Sun, located in the center of the M87 galaxy in the constellation Virgo, 50 million light-years (300 billion billion miles) away. This object's profile was revealed by the presence of colossal amounts of hot infalling gases, moving at enormous speeds into its dead center. Based on general relativistic principles, its Schwarzchild radius (distance from its event horizon to its central singularity) has been estimated to be several billion miles wide, about the distance from the Sun to the outer reaches of our solar system. An observer falling into such a dark cauldron would be "blessed" with many minutes of investigatory time, as he plunged into its dead center.

It would be cruel to contemplate a human explorer willingly entering such a sinister place solely for the purposes of taking astronomical measurements. So let us imagine that a ship bound for a black hole is operated by an intelligent computer. Named HAL by his designers, he is programmed to carry out such an investigation without flinching. A marvel of the modern age, HAL is equipped not only with special gravitational and spatial-temporal sensors, but also with the keen electronic mind and sharp analytical abilities required to process his experiences quickly and translate them into meaningful descriptions.

HAL relays his "personal" reports back to Earth by means of an experimental telepathic connection with another computer named JEANIE (built by Dixon enterprises). At the same time, a group of terrestrial astronomers, called the "ground team," trains an extraordinarily powerful telescope on the spacecraft, monitoring its progress toward its destination. Thus, in our tale, earthly observers are able to compare internal (transmitted by HAL) and external (seen by the telescope) accounts of the voyage. (Here, for the purposes of exposition,

Disk around a Black Hole in Galaxy NGC 7052
Hubble Space Telescope • Wide Field Planetary Camera 2

PRC98-22 • June 18, 1998 • ST ScI OPO • R. P. van der Marel (ST ScI), F. C. van den Bosch (University of Washington) and NASA

FIGURE 3.3   Shown in this photograph, taken by the Hubble Space Telescope, is matter falling into a large black hole near the center of the NGC 7052 galaxy. Astronomers have found that massive central black holes form a typical feature of galaxies (courtesy of NASA).

we are taking considerable liberties; in reality no one could immediately sense what is happening in a ship millions of light-years away. Such signals would take millions of years to reach Earth.)

Today is the day HAL's circuits have longed for, when they will finally test their mettle. At long last, HAL's ship, the *Kubrick,* has reached its destination: the central black hole in the M87 galaxy. HAL's sensing instruments detect extremely strong spatial curvature and exceptionally high radiation counts (from infalling matter) in the vicinity, two sure signs that their target is imminent. HAL announces back to Earth:

I am pleased to report that our mission is proceeding as scheduled. I estimate that the *Kubrick* will reach the event horizon of the M87 object within one hour. The ship's acceleration has been steadily rising and tidal forces on it are increasing, but so far all systems appear to be functioning normally.

Indeed, the ground team notices that the craft is moving steadily toward the vortex. The ship does not appear to be under excessive strain. Strangely, however, they discover that some of the ship's systems seem to be operating slower than usual. Because HAL does not report this slow-down, they assume it's a function of the gravitational time dilation induced by the black hole.

As the *Kubrick* cruises closer and closer to its target, the effects on it of spacetime curvature become increasingly apparent to earthbound onlookers. More and more, every aspect of its behavior seems to be crawling along at a sluggish pace, from its tediously turning clocks to its laggard emission of waste gases.

What effect, members of the ground team wonder, might this slow-down be having on HAL? Upon their request, JEANIE manages to transmit a message to him.

"We have noticed that your vessel appears to be functioning slower and slower. Does your photonic circuitry register this as well?"

"Neg-a-tive. I have not no-ticed such a phe-no-me-non." replies Hal, his words drawn out like treacle dripping from a spoon. Being a self-aware cybernetic entity, however, he is naturally concerned about possible malfunctions. To test matters, he begins to recite a poem taught to him when he was a young cyberling on Earth:

"Dai-sy dai-sy, give meee the an-swer dooo . . . "

At this point, the ground crew *knows* something bizarre is happening. HAL's transmitted words, once steady and strong, have slowed to a plankton's pace. Minutes, and then hours, have passed between utterances. They are still waiting for the second line of the poem.

Meanwhile, from Earth's perspective, a strange kind of deep freeze seems to have enveloped the ship. All of its functions seem to have come to a virtual standstill. Even the ship's clock appears to have radically slowed down and then stopped its ticking. Moreover, the craft doesn't even seem to be moving; rather it has halted in midspace, near the supposed periphery of the black hole.

The reason for HAL and his spaceship's snail-like approach to the M87 object pertains to the warping effects of black holes. Just outside of such a dark body's event horizon, spacetime curves so dramatically that the light cones of objects in that region tilt considerably, their time axes pointing at sharply different angles compared to those on Earth (or in another relatively "flat" sector of the cosmos). Thus, one

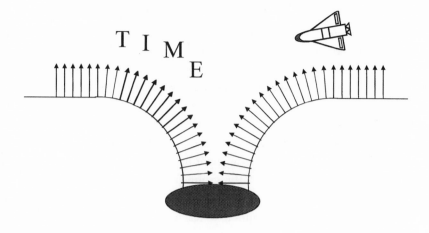

BLACK HOLE

FIGURE 3.4   Portrayed here is the distortion of time in the vicinity of a strong gravitational field—near a black hole, for example. A massive object bends both space and time in its vicinity, rendering time more "spacelike" and space more "timelike." Closer and closer to such a heavy body, the field gets stronger and stronger, causing temporal arrows to bend steeper and steeper.

second, from the vantage point of the ship while it passes through the "tilted time" region near the black hole, does not correspond to one second from a terrestrial perspective.

Once the ship has passed within the event horizon, this effect becomes even stranger. Its time axis has assumed such a great angle that the passage of indefinite amounts of ship time corresponds to zero seconds of Earth time. Therefore, no matter what transpires on the ship—for minutes, months, or millennia—outsiders observe it to be frozen in time.

One might understand this phenomenon by use of a simple analogy. Picture the Leaning Tower of Pisa, and imagine that a near-replica of it, called "the Standing Tower of Pisa," is erected right next to it. It possesses the same number of stories as the original, but is completely erect, pointing straight upward. These two towers, bent and upright, symbolize the axes of time in different regions of the cosmos—the former in a highly curved (strong gravity) part of space, such as near a black hole, and the latter in a relatively flat sector, such as on Earth.

We further assume in our analogy that for each tower, no matter what its inclination, it takes fifteen seconds to climb each story. Thus, it does not matter who is the climber and which tower is being scaled; the time to reach the top is the same. Since the Leaning and Standing Towers are each eight stories high, it takes two minutes to ascend each.

Now suppose two girls are playing a game, where they each have to climb up one of the towers to a height of 100 feet (as measured in the usual way, straight up from the ground), and drop a ball from the nearest window. The first girl climbs the Standing Tower, and notices that it takes her about one minute to reach the requisite altitude. She is about to drop the ball when she observes that her friend in the Leaning Tower hasn't reached the window yet. Perhaps her companion is walking more slowly? she wonders. Why is the second girl slowing down so much?

The answer is that by climbing the Leaning Tower, the second girl must traverse more stories, and hence take more time, to reach the same height as the first girl does. Although the first girl only needs to ascend four stories, the second, because of her tower's tilt, must climb five stories to achieve the same altitude above the ground. Hence the latter climber takes 25 percent longer than the former, experiencing a mock "time dilation." The more the Leaning Tower tilts, the more noticeable the discrepancy.

By analogy, those moving through a curved part of spacetime, where the time axes are tilted, tick off more seconds in reaching a goal than those watching from a flat region of spacetime would expect. Just outside a black hole's event horizon, this effect becomes highly pronounced; the "towers of time" lean considerably, angled more and more "horizontally" the closer one gets. At, and within, the event horizon itself, these "towers" are toppled so much that time runs in the normal direction of space—as different from ordinary temporal passage as a moving sidewalk is from an elevator.

Because, within a black hole's interior, time is spacelike, and space is timelike, some theorists have stated that these two dimensions "swap behaviors." In one sense, this statement is true; in another, it's not.

In the outside world, one's destiny normally follows an inevitable path toward the future; it moves one way in time. Within a black hole, on the other hand, one follows an irrevocable path toward the central singularity, where one is invariably crushed. Thus, spatial passage inside the event horizon assumes the inevitably of temporal passage outside it.

This situation is not wholly symmetric, however. In a Faustian exchange, by entering a black hole one sacrifices one's spatial freedom to gain a hollow sort of temporal liberty. In essence, one acquires the ultimately meaningless ability to move through external time as much as one pleases, as long as one remains within its twisted realm. One might journey to the thirtieth century, or the three-hundredth century, or the three-thousandth century of Earth time and gaze at light emanating from the far future. Into its escape-proof portal falls the entire sum of the radiation from all times to come, instantly providing glimpses at all ages of the universe-to-be. Hence, one might experience the whole of tomorrow in a flash.

But Mephistopheles would have the last laugh. Because light from the outside world would impact so rapidly—like a video run on fast forward billions of times quicker than normal—one might glean no meaningful information about the cosmic future. Moreover, one's personal time would still run mercilessly forward. One could not reverse the hands of the black hole's internal clocks. Therefore, the time travel one engaged in would be pointless; one would still be condemned to the singularity's crushing hell.

Accordingly, the fate of the spaceship *Kubrick* and its cybernetic inhabitant, HAL, does not seem particularly promising. Though outsiders picture them to be frozen at the M87 object's event horizon, their image slowly fading over time, in reality (according to their own time frame) they are plunging into the center of M87. Being a robot, HAL eschews self-pity and spends his final moments collecting and processing infalling radiation, gauging as much as he can from this data about the future of the universe. The task is essential hopeless; the light rains in far too fast. Just before his processors are ripped apart, however, he gains a glimmer of insight about the ultimate destiny of the cosmos—a thought that unfortunately he has no time to share.

One might wonder what would become of the material remains of those unlucky enough to be pulverized near a singularity. The immediate answer is that these would turn into pure energy, following Einstein's famous conversion principle. The energy would augment the mass of the black hole, and the entropy (amount of disorder) would increase its surface area. Then, by means of a mechanism known as Hawking radiation, this energy would slowly trickle back into space. For the average-size black hole, this radiative process would take trillions and trillions of years to complete.

Some theorists speculated that matter might escape from such objects through the "back door." Employing detailed mathematical constructions, they found hypothetical links between matter-gobbling black holes and postulated matter-gushing "white holes." Perhaps objects entering black holes might exit in another part of the universe by means of white holes, they pondered.

Further calculations, however, indicated that such connections would be highly unstable. An astronaut attempting to traverse such a shortcut would surely be crushed in the attempt. That is, if he wasn't already fried by the enormous radiative fields descending upon a black hole, or ripped apart by the tidal forces. In short, an attempted voyage through such a shrunken star would be agonizingly drawn out at worst, and mercilessly brief at best, in either case leading to death.

Expeditions to Tomorrow

In the late 1980s, Cal Tech theorist Kip Thorne, along with two of his graduate students, Michael Morris and Ulvi Yurtsever, worked out a model for "traversable wormholes": spacetime tunnels that superficially resemble black holes but allow safe transport to other parts of the universe. These cosmic connections, they found, could be produced by assembling a special hypothetical kind of negative-mass material, dubbed "exotic matter," into particular types of geometric configurations.

One unexpected consequence of wormhole dynamics stirred up a bustle of scientific controversy. Thorne and his students discovered that a special arrangement of the mouths (entranceways) of a wormhole would enable the possibility of backward time travel. In particular, if one mouth accelerates outward rather than inward relative to the other—like a boomerang sent out and then returning to its thrower—an astronaut might pass through the two portals in succession and leap back through time.

The way such an arrangement would work has to do with the well-known twin scenario of special relativity. In this thought experiment, we imagine that a roving astronaut has a brother the same age on Earth. If he blasts off in space, accelerates until he is traveling close to the speed of light, and then returns, Einstein's theory mandates that he would age less than his twin does. Depending on how fast the journey transpires, he could end up considerably younger.

A similar mechanism would guide a wormhole time machine. Suppose that timepieces are affixed just inside and just outside the two gateways of a traversable wormhole—four clocks in all. Because the two internal clocks (call them Inside One and Inside Two) are connected via a spatial shortcut, they must always maintain exactly the same time. The two external clocks (Outside One and Outside Two) would not be so constrained. Depending on the motion of the mouths they are near they might agree or disagree with their neighbors.

Now imagine that one of the mouths is stationary, and the other accelerates back and forth in the manner of the roving twin. For the static mouth, special relativity mandates that an observer situated within the wormhole entranceway would experience the same time intervals as a nonmoving onlooker located just beyond the portal. Therefore, the inside and outside clocks (Inside One and Outside One) would offer identical readings.

For the moving mouth, on the other hand, Einstein's theory demands that the internal clock (Inside Two) would run much slower than the external clock (Outside Two). Hence, upon the completion of the mouth's motion, the two would display much different times. Depending on how fast and how long the moving mouth travels before it returns to its original position, the difference could be minutes, months, or millennia.

Let us consider an illustrative situation. Suppose Sylvilla and Percilla are twin sisters in their late twenties. Sylvilla, the more adventurous of the two, volunteers to become the first time-traveling explorer, whereas Percilla remains behind on Earth. Sylvilla blasts off in space, then, after an uneventful journey, guides her ship toward the moving entranceway of the time hole. Before she enters the mouth, her ship's timepiece [synchronized with Outside Two] displays the year A.D. 3025. However, once she passes through the portal, it indicates that the year is A.D. 3000 [synchronized with Inside Two; also with Inside

One and Outside One]. She travels through the wormhole's "throat" and emerges from the stationary mouth in a different spatial region. Even after she is back in normal space, her clock [Outside One] continues to read A.D. 3000. When she finally lands on Earth again, she is shocked to find her sister barely a toddler. Her only challenge now is babysitting.

A similar theoretical time-hopping method, proposed by Princeton astrophysicist J. Richard Gott, involves objects known as cosmic strings: long, thin strands of concentrated energy that some astrophysicists believe were created in the early universe. These colossal, spaghettilike structures are thought to be so dense that a single inch of their material would weigh nearly 40 quadrillion tons—packing in quite a gravitational punch. Unlike wormholes, they would be made of conventional matter, albeit in unique arrangements.

Gott's time-travel model relies on the encounter of two cosmic strings, passing in opposite directions at near-light speeds. If passengers aboard a rocket ship headed toward the first string, whisked around it, and then managed to circumnavigate the second string as well, Gott calculated that general relativistic effects would permit them to travel backward through time. This time warp would result from the distorting influence of the strings' compact matter. The light cones of neighboring points near the strings would overlap with each other, allowing the past sections of each to contact the present sections of the next. In this manner, a so-called closed timelike curve (CTC) would be formed: a cosmic link between eras. The greater the gravitational impact of the strings, the more noticeable the distortion, and the more pronounced the leap back through time would be.

Gott's device possesses one special limitation. One could not use it to travel further back in time than when the stringy machine assembled itself or was built. So, unfortunately, unless we found an extant machine, constructed long ago, perhaps, by an intelligent alien race, we'd have little chance of using Gott's method to joust with King Arthur or take snapshots of the dinosaurs.

On the other hand, we could conceivably use cosmic strings or wormholes to enact voyages to the future and back. Let's consider the possible steps of such a round-trip journey. Traveling forward in time would be relatively straightforward, given the proper technology. We could simply hop on a spaceship moving sufficiently close to the speed of light and then return to Earth. Owing to the effects of special relativistic time dilation, we would then age at a much slower rate than those we left behind. In years, or even months, we could find ourselves experiencing the world many centuries ahead.

Who has not dreamed of traveling into the far future? Among the doting mothers and fathers, grandfathers and grandmothers on this planet, who would turn down an offer to find out what will happen to their more distant descendants? Innovative thinkers would welcome opportunities to discover the impact of their creative works upon generations to come. Inventive architects and engineers would cherish observing how their structures hold up throughout the eons. Political and corporate leaders would relish the chance to see what became of the institutions they headed. Beyond mere forecasting, through time travel one could *know* the future. As H. G. Wells suggested in his famous novella, time travel would represent the ultimate form of prediction—the ability to experience the world of tomorrow, then return and describe it.

After we have witnessed the future, we'd likely seek a way of journeying back to the present and reporting what we have found. Otherwise, we'd abandon our friends and family forever, and lose any chance to share our predictive knowledge. Perhaps we'd voyage homeward by means of a suitable wormhole or pair of cosmic strings. These would have to be constructed or found—not a trivial proposition, considering none have been discovered so far. Assuming we could find a way to return, armed with knowledge of the years ahead, we'd likely be hailed as veritable prophets.

But what if we used information from the future to change the present? Could we then alter the shape of things to come? In that case, which would be real, the future we've already witnessed, or the new time line we've created? If we've collected souvenirs of our journey, would they disappear once we've changed the conditions that made them possible? Or, conversely, would they remain as effects robbed of their original causes? For instance, if we brought back a work of art from the future, but then altered conditions so that the person who created it never became an artist, what would happen to the piece? If it vanished, where would it go? If it remained, who made it?

As theorists have noted, if CTCs exist in spacetime, generated through strings, wormholes, or some other mechanism, they would introduce a host of dilemmas regarding the nature of cause and effect. By traveling backward in time, one could infect earlier days with actualities yet to be, setting off a chain of events in the former that could alter the latter in contradictory ways. In such a manner, one could tamper with causality and engender innumerable paradoxes.

This Heading Is False

A paradox is a self-contradictory statement or argument. A classic example is the following phrase: "This sentence is false."

Let us consider the truth value of that statement—a typical case of what is called the Epimenides paradox. Following Boolean logic, it is either true or false. Neither conclusion, though, is self-consistent. Suppose the phrase is correct. Then, taking its words at face value, the sentence must be false. If it is false, however, then its meaning cannot be correct. Thus the sentence is true. Hence, the statement's truth value is indeterminate, a paradoxical situation.

Martin Gardner, in his influential "Mathematical Games" column, a staple for decades in *Scientific American,* has brought to his readers' attention a number of very curious prediction paradoxes. Two of the most fascinating of these are the "Paradox of the Unexpected Hanging" and "Newcomb's paradox." Before we examine the contradictions inherent in backward time travel, let's consider these examples of how even ordinary prediction might lead to baffling predicaments.

The "Paradox of the Unexpected Hanging," as Gardner points out, first came to light in the early 1940s, and formed the subject of dozens of scholarly articles in the 1950s and 1960s, including an incisive piece by Harvard logician W. V. Quine. Though many have written about the paradox, no one knows who originally thought of it. Here is the way it is usually stated:

> A judge condemns a man to be hanged, sentencing him on a Saturday. To make the man's punishment even more severe, the judge decides to obscure the date in which it will be carried out. "The hanging will take place at noon," he informs the prisoner, "on one of the seven days of next week. But you will not know which day it is until you are so informed on the morning of the day of the hanging."[3]

FIGURE 3.5   *Martin Gardner (1914–),
well-known popularizer of mathematics, introduced many curious puzzles and schemes in
his "Mathematical Games" column in* Scientific American *magazine, including John
Conway's Game of Life, the Paradox of the
Unexpected Hanging, Newcomb's paradox,
and various conundrums involving time
travel (courtesy of Martin Gardner).*

The condemned man turns pale. After all, this judge is known to be
a man of his word. Thoughts of doom spinning through his head, he
walks off with his lawyer for a private consultation.

As soon as they were alone, the lawyer begins to chuckle. The judge,
she informs her client, cannot possibly carry out the sentence. None of
the days in the week to come could provide possible times for the
hanging.

For example, she points out, suppose the judge waits until the following Saturday. Then, on Friday afternoon, her client would still be
alive. He would know with certainty that he would die the next day.

This would violate the edict about not knowing the date of hanging until the morning of the actual execution. Therefore, Saturday would be out.

That leaves Sunday through Friday. But if the judge chooses Friday, he would violate his own edict once again. The prisoner, alive on Thursday afternoon and knowing that Saturday has been ruled out, would realize almost twenty-four hours in advance that Friday would be his last day. Similarly, if the judge selects Thursday, the condemned man would know on Wednesday; if Wednesday, he would know on Tuesday; if Tuesday, then on Monday; if Monday, on Sunday. That leaves only Sunday as a possible execution day. But that can't be possible, because then the prisoner would be able to discern immediately when his last day on Earth would be. Hence, all of the days of the week are ruled out for the hanging.

Is the judge foolish, then, in making his stipulation about the element of surprise? Do his own words entrap him in a predicament with no resolution? The philosopher Donald John O'Connor at the University of Exeter, writing in the British journal *Mind* about a very similar predicament, argued that such a proposition cannot be carried out. Anyone issuing such a proclamation, he wrote, has established circumstances under which the action cannot take place at all.

Nonsense, rebutted another philosopher, Michael Scriven, writing in the same journal. The judge could carry out his sentence on any of the days—Thursday, for instance—and the condemned man would indeed be surprised. He would walk to the gallows cursing his lawyer and wondering why he ever trusted her convoluted thinking. As the prisoner was hanged, all would admit that the judge was a man of his word.

To resolve this paradox, one must realize that prediction is an individual action. A true prediction made by the judge might seem false to others up until the event takes place. To surprise the prisoner, the judge needed only to keep his plans secret. Just because the prisoner thought that he wouldn't be surprised doesn't mean that he wasn't. In summary, the prisoner's reasoning was faulty because he assumed that the judge would think the same way he did.

Another intriguing prediction paradox, proposed by physicist W. A. Newcomb in 1960 and first introduced to the general public by Gardner in the early 1970s, concerns the nature of free will. Newcomb's paradox imagines a situation in which somebody is asked to choose between two possible hidden prizes (as in the old American television game show *Let's Make a Deal*). The nature of that person's decision depends on his or her beliefs in predestination.

As described by Gardner, the rules of the game are as follows: "Two closed boxes, B1 and B2, are on a table. B1 contains $1,000. B2 contains either nothing or $1 million, You do not know which. You have an irrevocable choice between two actions:

1. Take what is in both boxes.
2. Take only what is in B2."[4]

If it weren't for an additional stipulation, the answer would be obvious: take both. However, to confound matters, sometime before you make your choice a superior, highly prescient Being (perhaps an ultraintelligent alien or a high-powered computer) predicts what you will do. Although he is not perfect in his forecasts, he is usually extremely accurate.

To confuse you he does the following: If he thinks that you will choose both boxes, he leaves B2 empty. If, on the other hand, he expects you to take only B2, he deposits $1 million there. In either case he leaves $1,000 in B1. Then, with a sly grin (or an electronic cackle, in the case of a computer) he lets you in on what he has done and challenges you to make the most lucrative decision.

If you were a strict Calvinist, and the Being were God, then the answer to the dilemma would be clear. Because you would expect him to be absolutely correct in his prediction, you had best pick only B2, and get your $1 million. Otherwise, he would have anticipated your decision to chose both and would have deposited nothing in B2. You'd be stuck with only $1,000.

Since the Being does very occasionally err, however, you may wish to consider the alternative. What if he logically deduces that you'd select B2, and places $1 million there? You trick him and choose both. Then you'd have an extra $1,000 to add to your million, and you'd be even richer. Conversely, what if the Being anticipates that you'd pick both, and deposits nothing in B2? In that case, if you select only B2, you'd be empty-handed. Maybe choosing both is not such a bad idea after all.

There is no right or wrong solution to Newcomb's paradox. One's instinctual response depends solely on one's belief system. The more one believes in free will, the likelier it is that one would choose both boxes, as that would offer $1,000 at worst and $1,001,000 at best. On the other hand, strict determinists would be more inclined to choose only B2, the option that the Being, if he guesses correctly, would bestow with the greatest reward.

One assumes in Newcomb's paradox, of course, that the Being does not have access to a time machine. If he were able to travel (or at least influence events) backward in time that would be a completely different story. Depending on the choice subsequently made, he could alter the contents of the boxes beforehand. The game, quite simply, would not longer be fair.

As Gardner and many others have pointed out, in general, the possibility of reversing present events through actions directed at the past would likely engender paradoxes far more outlandish than that suggested by the "Unexpected Hanging," or even by the diversion just described. Time-travel paradoxes offer snarled twists of nature's garment, so insidious that they cannot easily be undone. Examples of such contradictions offer bold challenges to those working in this highly theoretical field.

The Man Who Never Was

Tempus Zeitreiser III is paying a nostalgic visit to his hometown, thirty years before he was born. The way he pulls off such a trick has to do with a gleaming metal contraption, safely hidden behind a bush. You see, Zeitreiser is a highly inventive individual, for whom time is no obstacle (wink, wink). But let's keep his hobbies secret, as we don't want to blow his cover.

Tempus walks into a diner, rumored to be known for its exceptional milkshakes. He sits at the front counter and orders one—vanilla malt. "Mmm, delicious," he mutters. "It's a shame this place got knocked down . . . "

He stops himself, realizing that he ought not to say too much.

"What's that?" inquires a teenager sitting next to him.

"I mean, these milkshakes are so great, this diner will probably be famous forever."

Tempus strikes up a conversation with the teenager, whose demeanor and way of speaking seeming awfully familiar. "So what's your name, by the way," he asks the boy.

"Tempus Zeitreiser," responds the teenager.

"The first?"

"I guess so," says the boy. "I don't know any others."

"With a name like that," replies the visitor, "I'm sure you will go far. You could do almost anything."

"You think so?" answers the boy. "Say, I think you are absolutely right. I was just about to propose to my girlfriend, Peggy Sue. We'd get married after high school. But I'm sure I can do much, much better than that."

"Better than grandma?" thinks Tempus III.

"So that's that," says Tempus I. "I'm just not going to get married now, perhaps not for another twenty or thirty years or so. I'm going to break off with Peggy Sue right away and see the world instead. Thanks, sir, for your advice."

But, having no actual grandpa, no parents, and no record of his birth, the visitor from the future vanishes, a victim of the infamous "Grandfather paradox." Because his grandparents never got married, and never had children, let alone grandchildren, he simply cannot exist.

If the situation resolves itself in such a clear manner, with the nonexistence of the perpetrator, then why it is a paradox? Well, it doesn't. In the absence of the time traveler, young Tempus I certainly would have married Peggy Sue. They certainly would have had children and grandchildren, including Tempus III. But then, Tempus III would have taken up time travel and performed his foolish excursion, blocking, once again, his grandparents' marriage. Thus, reality contradicts itself in an irreconcilable set of circumstances—a true Gordian knot.

Some theorists—for instance, Roman Jackiw of MIT and Stanley Deser of Brandeis—argue that the existence of the Grandfather paradox negates the possibility of time travel. Causality, they attest, is such a vital aspect of physical law that it cannot be violated. Therefore, any conceivable mechanism that might tamper with the order of cause and effect must be considered unnatural.

To counter such objections, Russian scientist Igor Novikov, along with Thorne, Morris, Yurtsever, and several other physicists, have postulated a new law of nature called the principle of self-consistency (PSC). This hypothesized prescript states that all solutions of Einstein's equations and other physical formulae must be internally consistent for all times. That is, a system's parameters must always be unequivocal, never harboring any twists or paradoxes.

Within the strict limits prescribed by the PSC, self-consistent time travel scenarios would be possible, these researchers assert. A voyager would be able to return to the past as long as he either made no significant changes—or, if he did alter something, it was destined to be altered anyway. For instance, if a time traveler stole the *Mona Lisa* from the Louvre and brought it back to da Vinci's studio, then convinced him to adopt it as his own work, history's cohesion would be maintained. Because none, save the culprits, would ever know that da Vinci never painted his most famous work, no logical contradiction would be engendered.

Such neatly tied packages, however, offer paradoxes of a different sort. Rather than history vacillating between two contradictory approaches, no discrepancy exists. Yet something is still missing: the act of creation. Who, in the da Vinci example, actually painted the *Mona Lisa*? Transferred from the Renaissance master's studio to the Louvre, and later brought back from the Louvre to the studio, from whose brushstrokes did art's most famous smile originate? Strangely enough, the answer would be that the painting created itself.

Normally, one supposes that one cannot fashion something out of nothing. Nonetheless, backward time travel suggests the opposite— that self-created objects are fine as long as they are self-consistent. In the manner of a magician, pulling a rabbit out of a hat would be an easy trick; just send a floppy-eared creature from the future back in time and have it materialize in the past. Then make sure to use the same bunny the next time around. It wouldn't have a mother or a father; it would just *be*.

There is one final paradox to mention—probably the hardest to resolve. This quandary, which I call the "changing one's mind" paradox, is closely related to the "something out of nothing" paradox just described. It involves the situation in which someone with a time machine makes the conscious decision to use it later to bring something back from the future to the present, then changes his mind after the item has already been sent.

Consider, for instance, the case of Ivana Dolittle, the twenty-third century's most spoiled heiress. Among the riches she has inherited from her wealthy father is a time machine of his own design. Though in his will he advised her to use it only when absolutely necessary, she has thrown all caution to the wind, and has come to depend on it to satisfy her every need.

One afternoon, just before she is about to head out for the annual F. Scott Fitzgerald memorial ball, she spills brandy on her best dress. Normally, she'd simply change outfits, but for that event she has her heart set on wearing it. To remedy the situation, she decides to have the dress sent out for dry cleaning after she gets back from the ball. Once it is ready, she'll send it back in time, programming her time machine for the date of the ball. Sure enough, once she makes the decision to do so, a clean dress "magically" appears right next to her.

Everything goes splendidly at the ball. A handsome young website designer, who has made a fortune buying and selling Internet stock, sweeps her off her feet. Soon, she is on a one-way flight with him to Rio de Janiero, where they get married. Madly in love, she forgets all about time travel, and never sends the dress off to be cleaned. She lives happily ever after within her fabulous world of wealth and paradox.

Who cleaned the dress? In this scenario, not only is there a discrepancy between cause and effect, even worse, there is an effect without a cause at all. Either the dress appeared completely out of the blue, or, somehow, somebody else performed the task.

To restore self-consistency to the tale, one would have to invent another possible way the dress was cleaned and sent back in time. Perhaps Ivana's father anticipated her headstrong behavior and built in an automatic device that would ensure that any intended decision is carried out. The device, then, phoned the dry cleaner and had the dress picked up, cleaned, and returned. Then it sent the dress back in time, to carry out Ivana's intention. Such a device, however, would have had to have read her mind—an unlikely proposition.

Apparently, the existence of free will creates an enormous stumbling block for backward time travel. In a wholly deterministic world, time travel would stand a chance of being self-consistent. Considering, though, the seemingly impulsive nature of much of human decision making, backward time travel would be problematic at best.

The Chronology Protection Conjecture

Most physicists, because of their training, would prefer to settle the matter of backward time travel by means of mathematical, rather than philosophical arguments. Whether past-directed leaps are possible or impossible, they feel, let the equations decide. Classical general relativity seems to allow these jumps, but it does not represent the whole picture. The true resolution of the question would likely derive from a complete reading of quantum gravity.

Quantum gravity is the idea that general relativity represents a classical approximation of a fully quantized theory of gravitational fields. On the tiniest levels, it suggests, the universe harbors probabilistic elements—minute variations in the topology of spacetime known as quantum fluctuations. Though theorists have yet to develop a model successfully combining general relativity and quantum mechanics, they have made considerable progress. Some believe that they understand its limits well enough to rule on the time-travel question. Therefore, a number of recent publications have attempted to settle the matter.

Rallying the anti-time-travel camp is an influential figure in cosmology, none other than Stephen Hawking. In a bold critique of contemporary time-machine schemes, called the "Chronology Protection Conjecture," he states unequivocally: "The laws of physics do not allow the appearance of closed timelike curves."[5]

To prove his assertion, Hawking calculates how spacetime might behave under a variety of potentially causality-violating conditions. He finds that in many cases the structure of reality resists the formation of CTCs. Often, if one tries to form a CTC, the region of space in question closes itself off and becomes wholly inaccessible. Perhaps, Hawking suggests, that is nature's way of protecting its history.

Hawking concludes his paper with a more mundane argument against time travel: "There is also strong experimental evidence in favor of the conjecture from the fact that we have not been invaded by hordes of tourists from the future."[6]

Another crusader on Hawking's side is Washington University professor Matt Visser. Visser, who is enthusiastic about the possibility of traversable wormholes, and has even designed a few models himself, realizes that their conceivable use in time travel cools off some to the idea. He points out that if the chronology protection conjecture were true, critics would have less ammunition against the wormhole idea. Indeed, in his research on quantum gravitational effects, he finds that a "gravitational back reaction" blocks travelers from using wormholes to pursue their pasts. Like great waves preventing a small boat from docking on a rocky coast, quantum fluctuations in the metric (web of connections) of spacetime, according to Visser, would stop explorers from entering such forbidden zones.

But certainly not all scientists agree with Hawking and Visser's conclusions. Leading the pro-time-travel defense is J. Richard Gott—who had apparently hoped that his original cosmic string proposal would have wrapped up the whole argument. Unfortunately, as critics pointed out, at the very least, it left a number of loose ends. Undaunted by these critiques of his results, he has continued to emphasize that the work of Hawking, Visser, and others does not rule out all time travel possibilities, only some of the cases. In rebuttal, Gott, Thorne, and other researchers have discovered many counterexamples—situations in which CTCs are conceivable. To consider the chronology protection argument physical law, sufficient reason must be found to dismiss all possible exceptions. Gott and his colleagues are not satisfied that this is the case. As Gott has written:

New objections to spacetimes with CTCs can continue to surface, as old problems are put to rest, so it might seem that disproving the chronology protection conjecture would be a tall order. But, proving there are no exceptions to the chronology protection conjecture, ever, would seem to be a daunting task. This is particularly true since we currently do not have a theory of quantum gravity or a theory-of-everything.[7]

Indeed, Gott appears to be correct on this point. Until quantum gravity is fully understood, we cannot unequivocally rule out the mere possibility of time travel. To say it is unfeasible—unlikely even—is one thing, but to declare it impossible for all times requires a higher level of proof than theorists might currently provide.

One might wonder why leading physicists have spent valuable time debating a subject as speculative as time travel. To unravel the issue of prediction and causality in general relativity, they realize, science must resolve certain fundamental questions about its structure. What are the limitations to spacetime topology? Can closed timelike curves exist in nature? If so, is physics still self-consistent, or does it harbor discrepant paths in time? In describing the microscopic dynamics of spacetime, researchers hope that a complete quantized theory of gravitation will answer some of these quandaries.

Uniting relativity and quantum physics would complete the revolutionary process begun in the opening decades of the twentieth century: the toppling of Laplacean determinism. In the 1800s, science spoke confidently about the possibility of knowing the future for all times. Although Minkowski's model of spacetime seemingly cemented that picture, Einstein's general relativistic proposal—embracing myriad complex geometries—opened up a lot of cracks. Quantum theory, developed shortly thereafter, shattered mechanistic concepts even further by bringing probabilistic dynamics into physics. Prediction, it suggested, possessed absolute limits.

Fittingly, these radical changes began during an age of utter chaos in Europe. When uncertainty was embraced by science, it was an era of falling empires, shifting alliances, and senseless destruction. In light of the unanticipated horrors of World War I and the unforeseen impact of its mosaic of consequences, henceforth who could reasonably claim the direct path to future knowledge?

4

ROLLS OF THE DICE

The Quantum Future

In those halcyon days I believed that the source of enigma was stupidity. But now I have come to believe that the whole world is an enigma, a harmless enigma that is made terrible by our own mad attempt to interpret it as though it had underlying truth.

—UMBERTO ECO
(Foucault's Pendulum),
trans. William Weaver

It is very hard to predict—especially the future.

—Saying attributed to
quantum theorist NIELS BOHR

Age of Uncertainty

A majestic baroque church constructed, stone by stone, over many decades—silently standing guard over a humble village for countless centuries—utterly destroyed in a matter of hours. A once-bustling city of meticulously planned broad boulevards, lined with cafés and news-stands, speckled with statues and fountains, now a heap of crumbling ruins—blackened visages painted with the ashes of nightmares. Boy soldiers, with eyes full of hope, piled motionless in vast trenches.

Perhaps no interludes in history greater exemplify the fleeting nature of fortune's warm embrace than the cold, dark days of armed conflict. When one considers contemporary warfare's capacity for instant, massive destruction, one shudders to think of so many lives changed forever by senseless deaths of family and friends. In the case of World War I, the first modern war, the shock of unprecedented casualties and destruction added to the enormity of the despair.

The destruction of towns and villages in the Great War heralded an even more fundamental change in the European way of life. Once mighty, the Russian, German, and Austro-Hungarian empires collapsed, leaving chaos in their wake. Hunger, homelessness, unemployment, and inflation besieged much of Europe for many years.

Not all of this was sudden. Before the war, there were, without a doubt, many omens of ill times to come. Throughout the late nineteenth and very early twentieth centuries, local conflicts and civil unrest, including assassinations, strikes, and attempted coups, intermittently plagued Europe and other parts of the world. But in spite of the warning signs before the war, no one expected the severity of the turmoil that would engulf many nations in the 1910s.

The turbulence of European life inspired a different sort of literary and artistic chaos. In cafés, bookshops, galleries, and theaters, the dadaist and expressionist movements flourished, emphasizing a new sort of art and literature devoted to reflecting the random and spontaneous nature of the real world.

In dada, simple household items such as chairs or bicycle wheels were put forth as objects of art. Dadaist films emphasized random meaningless events and kaleidoscopic displays of imagery. Similarly, the literature of this genre played with nonsense passages, stream-of-consciousness exposition, and, most of all, a display of wild humor aimed at the uncertainty of life.

One of the principal writers of this period, living in Zurich, was the self-exiled Irishman James Joyce. Joyce pioneered a stream-of-consciousness style of writing that attempted to incorporate fragments of the thought processes of the characters depicted in his stories. In his work, Joyce created a potpourri of images, emphasizing the direct sensory experiences of those he wrote about and leaving the interpretation of these scattered visions to the reader. Joyce's approach derived in part from a growing realization that mental processes take place in layers of thought and feeling, with much occurring beneath the surface. More than by anyone else, this realization was spurred by the work of Freud, who invented a new lexicon of the mind. Consequently, as Joyce expressed in his writings, the linear model of human thought was replaced by one resembling a labyrinth.

The late teens and twenties, especially in European cities such as Zurich, Paris, Vienna, and Berlin, brought forth a clearinghouse of seemingly bizarre ideas in art, music, and literature, with each area of the fine arts challenging the others to push their limits further. Although dada reigned supreme in many of the cafés and bookshops, jazz, with its exotic mélange of melodies, found eager audiences in smoky bars. Offering visual counterparts to jazz, cubist and surrealist art presented brilliant colors, arresting geometric forms, and strangely assembled images. Postwar Europe was truly reveling in randomness.

While this revolution in art, literature, and music was taking place in the coffeehouses of Europe, an equally turbulent revolution was taking place in physics. Radical new concepts, shattering the very framework of this science, captured the public imagination in an unprecedented way. We have examined how relativity, the first twentieth-century revolution in physics, rendered space and time a complex web. Quantum physics, the second revolution, ensured that the bulk of the passages of that maze shall remain hidden from view, accessible only to those who chance to stumble on them. Although relativity expanded the scientific lexicon to include the effects of observers' velocity, acceleration, and gravity on measurement—requiring two onlookers, under certain circumstances, to obtain two different readings of a physical parameter such as duration—quantum theory demonstrated even subtler relationships between how someone measures a quantity and what results he or she gets.

In considering a scientific prediction, quantum theory required one to examine the steps taken by the predictive scientist. In measuring two properties, changing the order of observation might wholly alter the results. Observing one quantity with great precision might make it impossible to find out anything at all about another. In short, although classical physics excluded the influence of observer upon observation, quantum theory mandated that such effects must always be taken into account.

If that were all quantum theory said about prediction, those of a deterministic bent would have been able to adjust. Scientists typically keep careful records of what methods they use. By accounting for the influences of these procedures, determinists would have eventually been able to anticipate how each experimental step affected the data. Exact predictions simply would have required a bit more work.

But the radical perspective of quantum physics offered even steeper hurdles to forecasting the future. Much to the horror of "traditionalists" such as Einstein—who, on this matter, stood fast—quantum theory injected hefty doses of randomness into the veins of physical science. No longer could researchers maintain the conceit that they could theoretically predict with great certainty the outcome of any natural process. Rather, in many cases, particularly involving atomic and subatomic processes, they needed to satisfy themselves with knowing only the probabilities of particular occurrences. Thanks to the work of quantum pioneers such as Bohr, de Broglie, Schrödinger, and especially Heisenberg, uncertainty came to be seen as a fundamental property of nature.

German physicist Werner Heisenberg first began his study of the world of the atom in 1919, the same year that Einstein's theory of general relativity was first publicized to a mass audience. As leftist revolution raged in the background—centered in his native Munich—Heisenberg set out on his own radical course: to construct a self-consistent model of atomic processes. By the mid-1920s, he completed his theory of "quantum mechanics."

Though Heisenberg originally developed his model to explain the orbital and transitional properties of electrons in atoms, in the years to follow, other quantum theorists enlarged the scope of its applications. Eventually quantum theory came to encompass a range of phenomena broad enough to include the mundane as well as the extraordinary, from computer chips and transistors to black holes and the very early universe itself.

In proposing his theory, Heisenberg wanted to explain how electrons orbiting atomic nuclei suddenly give off photons and migrate to lower orbits. Though the effects of such transitions could be seen in colorful spectral lines, each radiating in the characteristic frequency of an emitted photon, no one could adequately describe their underlying mechanism. Danish physicist Niels Bohr had offered a model with

FIGURE 4.1 *Werner Heisenberg (1901–1976), discoverer of the Uncertainty principle, one of the most important limitations to prediction (courtesy of Max-Planck-Institute and the AIP Emilio Segre Visual Archives).*

predictive success, but lacking deeper explanation—like an algorithm predicting incidence of sunspots without elucidating their cause. The German theorist wanted to do much better, and he did.

Heisenberg's own version of quantum mechanics involved a process known as matrix multiplication: a form of multiplying and adding terms in a series of tables. For each matrix, he represented initial states of electrons as rows, and final states as columns. Consequently, each transition, from one state to another, corresponded to a single matrix element (entry in the table). Much to his delight, he discovered that he could represent simple atomic processes by constructing equations involving such matrices. He then found that he could manipulate these equations to solve for quantities such as the momenta and positions of electrons. Heisenberg called these equations "quantum-mechanical series."

Heisenberg soon discovered that his formalism contained a peculiar property: one cannot solve for the exact position and exact momentum of a particle at the same time. If one specifies the position of a particle, one cannot know the momentum with any certainty; rather, one must reckon with a set of probabilities that the momentum is a certain amount. The same thing holds true if one knows the momentum exactly. In that case, the position cannot be exactly known. Furthermore, the imprecision in knowledge of the position is *inversely proportional* to the imprecision in knowledge of the momentum of a particle. The more precise one value is known, the less the other can be determined. This severe limitation on scientific knowledge, particularly on the atomic level, became known as Heisenberg's uncertainty principle.

The uncertainty principle might be envisioned as a television screen with two controls: one adjusting the clarity of the vertical, and the other, of the horizontal. Unfortunately, a prankster working at the factory that manufactured the set rigged it so that only one dial can work at a time. If one tries to steady the vertical, images start to leap left and right, until everything becomes smeared horizontally. Conversely, if one tries to adjust the horizontal, everything on the screen begins to move up and down—becoming a vertical blur. Under no circumstances might one focus the picture in both directions. Similarly, according to Heisenberg's prescript, physicists cannot simultaneously "focus" both the locations and momenta of particles.

An equivalent, but more intuitive, way of representing this principle and of understanding quantum theory was developed by the physicists Louis de Broglie and Erwin Schrödinger during the mid-1920s. De Broglie, in his Ph.D. thesis, put forth the notion that matter, like light energy, has wavelike properties. He reasoned that since light has wavelike attributes (it has a frequency and an amplitude) and particlelike attributes (it is formed of photon particles), matter should also exhibit this duality. Thus, all objects, from electrons to tennis rackets, should have a characteristic frequency, oscillating at a certain rate per second.

De Broglie's bold hypothesis proved to be a smashing success in predicting the behavior of electrons in atoms. It meant, however, that one could no longer think of electrons or other objects as having fixed positions in space. Rather, they must be thought of as probability waves, amorphous objects having locations spread out over regions of space.

These strange entities, displaying wavelike as well as particlelike characteristics, came to be known as wavefunctions.

With electrons and other subatomic particles successfully described as wavefunctions, quantum theorists sought an equation to predict the motion and spread of these probability waves. That task was left to Schrödinger, who brilliantly produced a wave equation for matter. Schrödinger's equation has been shown to be completely equivalent to Heisenberg's formalism—accurately reproducing the uncertainty principle.

The Nobel Prize–winning work of Heisenberg, de Broglie, and Schrödinger has formed the basis of modern quantum mechanics. It has led to a host of scientific discoveries that have paved the way for countless technological breakthroughs. In spite of the theory's marked success, though, its philosophical basis makes many feel uncomfortable. One has to "lose one's common sense in order to perceive what is happening," as Richard Feynman puts it. This is particularly true when one considers the mystery of quantum collapse.

The Mystery of Quantum Collapse

Like many a young lover, quantum mechanics is deterministic in its head, but random in its heart. Schrödinger's equation operates in as mechanical and consistent a manner as one would hope. It precisely predicts the behavior of wavefunctions—determining their forms indefinitely far into the future. In principle, given the wavefunction for an electron in A.D. 2000, and given all of the forces on that particle, one might accurately forecast the wavefunction's appearance in A.D. 3000—assuming nobody tries to measure it.

Ah, but there's the catch. Wavefunctions are extremely temperamental. Any measurement of their properties sends them into states of collapse. Not nervous collapse, mind you, but a curious random reduction that depends on what physical aspect one is observing. For example, if a physicist measures the position of an electron, the electron's wavefunction folds up like a pressed accordion into one of its position eigenstates.

Eigenstates represent projections (shadows) of wavefunction monoliths, cast in a manner that depends on the quantity observed. These are produced by applying quantum operators (functions corresponding to various physical aspects; position operators; momentum operators; and so on) to wavefunctions. As in the case of the Moon, car headlights, and streetlamps each bathing a pillar in light and forming distinct kinds of shadows, various quantum operators project the same wavefunction onto different sets of eigenstates.

Undisturbed, wavefunctions comprise linear (simply additive) combinations of eigenstates. Somehow, once a scientist takes a measurement, quantum dice are rolled, and the wavefunction instantly collapses into one of its random components—chosen on the basis of a well-defined probability distribution. The result he or she would obtain for the observation in question would be the eigenvalue (special number) corresponding to the particular eigenstate that chance has selected. For instance, measuring the momentum of a proton would force it into one of its momentum eigenstates; the actual momentum perceived would be the corresponding eigenvalue.

This standard way of understanding how wavefunctions collapse was developed by Niels Bohr. In honor of his native city, it is known as the Copenhagen interpretation. Though this explanation is successful, it remains controversial, because unlike any other known laws of science, it includes the will of the experimenter as an essential part of experiments. Until an observer measures something a wavefunction is intact; upon measurement, it instantly collapses. Somehow the act of looking plays the dynamic role normally assigned to forces.

The James Randis of physics have always been skeptical about this assertion. A purported psychic would be laughed out of a laboratory if he claimed that his extrasensory abilities only worked if he wasn't being tested, because the act of measurement adversely affected his powers. Yet, clashing with the time-honored scientific tradition of complete objectivity, according to the Copenhagen interpretation of quantum mechanics, scientists critically influence the nature of the experimental results they obtain. Moreover, the product of this influence—quantum collapse—is a fundamentally random process. Sharply contrasting with gut-felt notions about nature, no wonder many traditional thinkers have been troubled by this reading of quantum theory.

The famous "Schrödinger's cat" thought experiment nicely illustrates some of the disconcerting aspects of the Copenhagen interpretation. Imagine a cat placed in a closed black box such that nobody can see it. The cat is hooked up to an intravenous device designed to deliver a lethal injection of the drug "Cat-A-Tonic." The drug automatically flows if and only if a specific experimental condition is met. This criterion concerns a property, called spin, of an electron specially placed inside. If the spin is one way, the cat must die; the other way, it is spared.

According to quantum theory, electrons possess two possible spin states: up (counterclockwise) and down (clockwise). Until an observer measures an electron's spin, it is said to comprise an equal mixture of the two states. The instant he or she tries to obtain its value, its wave function collapses into either the spin up or spin down eigenstates, with equal likelihood of each. One of these eigenstates means that the cat dies; the other, that it remains alive. We suppose, however, that no observer measures the electron's spin directly; the only effect it has is upon the cat's state.

The dilemma posed by this thought experiment concerns the condition of the cat until its fate (along with the electron's spin state) is revealed by opening the box. Clearly, the cat's life or death depends on the electron's ultimate spin state. Until a measurement is taken, however, the electron resides in mixed state. Therefore, the cat—hooked up as it is to the electron—must also exist in a "mixed state" between living and deceased. The cat remains a half alive/half dead "zombie" until the point that an observer opens the box and the cat and electron's wavefunctions collapse in tandem. In short, the act of scientific measurement instantly changes a cat from a "mixed-state zombie" to either a living or dead animal. One can see why the Copenhagen interpretation gives even its ardent supporters some pause.

Despite reservations about the philosophical underpinnings of the Copenhagen interpretation, few scientists see reason to abandon it. Although, throughout the years, various theorists have developed potential alternatives, none have proven both accurate and successful. Many theories that purport to explain quantum phenomena in a simpler, observer-independent manner have since been proven incorrect. For example, a variety of tests have revealed that the so-called "hidden variable" approaches, in which unseen parameters guide quantum collapse in a wholly deterministic fashion, constitute nothing more than wishful thinking. Dismissing the chance elements of the Copenhagen interpretation, Einstein himself was a proponent of such mechanistic models, and was deeply disappointed that none in his day won legitimacy.

Recently, physicist Mark Hadley of the University of Warwick in England has revived this debate by turning to another of Einstein's ideas: the notion that elementary particles comprise miniscule warps in spacetime. In Hadley's model for quantum state reduction, spacetime knots called "geons" gravitationally interact on a fundamental level to twist and distort each other. Wavefunction collapse occurs when such interactive effects cause the topology of a geon to change from one form (the complete wavefunction) to another (the reduced wavefunction).

Another idea, called the spontaneous localization (SL) approach, advanced by Philip Pearle of Hamilton College and developed along somewhat different lines by Italian theorists GianCarlo Ghirardi, Alberto Rimini, and Tulio Weber, involves modifying Schrödinger's equation through the addition of an extra term. This new factor generates minute fluctuations in the evolution of waves—initially undetectable variations that ultimately cascade into full scale collapse. The entire process is akin to pebbles being removed from a mountain wall, disrupting larger stones and ultimately causing an avalanche. Therefore, even though this theory contains a small measure of randomness, the capriciousness of chance applies only on the particle scale, not on the human level.

Yet another alternative, called the decoherent histories (DH) approach, was suggested in its original form by Robert Griffiths in 1984, independently proposed by Roland Omnes, and later rediscovered by Murray Gell-Mann and James Hartle, who refined the theory and gave it its name. This interesting model walks a tightrope between classical determinism and quantum probability. It marks a return to the consideration of traditional observables (position, velocity, and so on) instead of wavefunctions as the basic descriptors of physical reality.

If, according to the Copenhagen approach, the scientific observer is king, the DH approach effectively dethrones him. Instead of concerning itself with the immediate effects of measurements, it emphasizes the larger picture; the universe itself is its domain. This is achieved by ignoring small fluctuations and seemingly random trajectories through a process called coarse-graining.

Coarse-graining is analogous to what an art critic does if she is standing too close to a painting in a museum, so near that she sees nothing but seemingly haphazard brushstrokes. She steps a few feet back, and takes in a larger view of the canvas. What she first viewed as a mere jumble now comes into focus as an evocative piece of art—possessing a unmistakable pattern.

Similarly, the DH method looks at quantum histories on a scale in which they appear deterministic and independent. If on one level the trajectories of particles appear wholly random, like the aberrant motions of novice drivers, this approach advises examining the scene by means of a wider-view lens. This greater perspective is achieved by assigning probabilities to each sequence of jumps, and then performing a sort of average of these possibilities to discern overall behavior. Ultimately, the DH approach asks us to understand physics as a collection of consistent paths that, on the largest scale, reproduce known classical behavior.

In terms of describing nature, these varied suggestions are different in spirit, but equivalent in accuracy, to the standard approach. Nevertheless, none so far has generated the general support needed to topple the leading depiction of collapse. Tradition seems to favor Bohr's venerable account—even in the face of many intriguing alternate possibilities.

Perhaps the most famous (and most radical) alternative to the Copenhagen approach is the intricate model known as the "Many-Worlds interpretation." Speculating about the division of wavefunctions into myriad near-identical images, it offers quantum mechanics a novel metaphor—ever-reflecting halls of mirrors instead of ever-collapsing houses of cards.

The Roads All Taken

In 1957, Princeton graduate student Hugh Everett, in a startling and original doctoral thesis, advanced the Many-Worlds interpretation as a possible alternative to the Copenhagen interpretation. According to his novel theory, wavefunctions never collapse. Instead, whenever an experimenter measures a particular quantity—position, for instance— of a given particle, the universe bifurcates into countless branches, each corresponding to one of the possible results of that measurement. In each fork of the universe resides a different eigenstate of the original wavefunction, pertaining to a separate result of the measurement taken. For example, the particle might be located, in alternative realities, either three-, four-, or five-billionths of an inch from the center of an atom.

Reacting to Everett's thesis, critics argued that no such splitting of the universe had ever been observed. Everett's response was to point out that no one would ever notice that reality had forked. Instead, like a gondola floating quietly during a nighttime tour of the canals of Venice, actuality would flow effortlessly into one passage or another, its passengers never noticing the other channels. Along each path, wavefunctions would perfectly obey Schrödinger's equation in a wholly deterministic fashion—with each transition absolutely seamless. In short, according to Everett, no conceivable experiment would be able to show that the other realities existed.

Consider how the Many-Worlds model eliminates some of the uncomfortable aspects of Schrödinger's thought experiment. Instead of the cat existing in a half-living, half-dead state until observation forces it to collapse, its reality simply branches. One of the forks corresponds to feline life, and the other, to feline death. The two alternatives each harbor a separate version of the experimenter, one rejoicing and the other in mourning when the box is opened. Because neither copy would ever know about the other, no contradiction would offer them cause for alarm.

Though Everett's hypothesis resolves some aspects of the collapse dilemma, it raises far more questions than it answers. Why would reality perpetually divide into innumerable, near-identical copies of itself? Exactly what is it about the act of human observation that would foster such bifurcation? Moreover, what principle of physics would prevent observers from contacting alternative universes? In short, to unload the burden of observer-induced collapse, must one take on the even-heftier weight of myriad ever-replicating cosmos?

The qualitative difference between classical physics, the Copenhagen interpretation of quantum theory and the Many-Worlds model can be illustrated by a simple example. Picture a forest with two houses, one owned by Mr. Hawk and the other by Ms. Dove. Many twisted paths lead from one house to the other, along which the postal worker, Carrie R. Pigeon, might travel when she stops to deliver letters.

One day, Dr. Stoolpigeon, a member of the employee efficiency unit of the post office, suspecting that his colleague is not performing at her peak, decides to find out if she is walking briskly and taking the best possible route between the houses. Speaking with his superiors, he requests permission to use a "classical technique" to do the job—namely, spying on her every movement. He wishes to follow her closely, as she embarks on her journey from Hawk's to Dove's residences, recording her position and velocity at every point.

Unfortunately for the nosy doctor, a new "uncertainty edict" precludes him from such close monitoring of Carrie. Constrained to a more limited search—of either her location or her speed, but not both—he elects to hover around the middle of the forest and check on which path she is traveling. Sure enough, he observes her walking along the longest, twistiest trail. (In conventional quantum terminology, her wavefunction has collapsed to that particular eigenstate.) He submits a highly negative report to his superiors, condemning her for selecting the least efficient route.

"My dear deluded Dr. Stoolpigeon," begins the response to his report. "You must have been dreaming too much about Copenhagen summers. Our 'Many-Worlds' contract with the postal workers specifically forbids us from judging a worker's performance on the basis of a single examination. In the spectrum of possible realities, each representing a different path she might have taken, you have singled out but one. Other Stoolpigeons, in other realms, would have observed complete different walking patterns—the shortest route, for instance—and arrived at wholly different conclusions. You must not condemn anyone for such a mere accident of fate. Cluck, cluck, you should be ashamed of yourself. Yours, Dr. Eaglet."

Island Instants

Oxford professor of quantum computation David Deutsch has recently proposed a variation of Everett's scheme that offers a fascinating possible solution to some of the most troubling conundrums involving the nature of cause and effect (the time travel paradoxes discussed in the previous chapter, for instance). In spirit, Deutsch's approach resembles, in particular, the strands of the Many-Worlds model stitched into the tapestry of an eternal block universe. The elegant cosmic fabric he has fashioned—a sort of megablock universe called the multiverse—is so tightly knit that causality paradoxes apparently cannot rip it apart.

The multiverse, according to Deutsch, harbors all possible worlds that obey the laws of physics. So, for instance, because physics allows atomic electrons to spin either up or down, with 50 percent likelihood for each (unless the spins of the other atomic electrons are already known), the multiverse embodies these options in equal proportions. That is, for a given unidentified electron, in half of all the component universes that form the multiverse it would spin up and in half it would spin down. However, in no sector of the multiverse would an electron possess a charge different than its normal amount, because physics would not allow that. Nor would one expect to find regions in which gravity sends apples flying up from the ground into trees either. Only physically consistent realms form the multiverse, and those in proportion to their probability of occurrence.

As in the case of Everett's model, Deutsch's approach avoids the need for wavefunction collapse in quantum physics, and thus circumvents uncomfortable situations such as Schrödinger's cat dilemma. Whenever physical law permits a particle to have more than one possible state, the multiverse partitions itself into these options (or, more accurately, it is already partitioned, since in Deutsch's system time does not flow). What the Copenhagen interpretation would describe as the chance for a wavefunction to collapse to a certain state, Deutsch's model would depict as the percentage of worlds in the multiverse with that particular condition.

Hence, if an experimentalist is measuring the position of a proton, and quantum theory assigns a probability of 5 percent that it is more than one inch to the right of a particular marker, then 5 percent of all regions of the multiverse associated with that instant would assign that proton to be more than an inch to the right. The other 95 percent of the sectors would be virtually identical, except that the proton would be less than one inch to the right. The experimentalist would determine which sector he is in once he completes the measurement.

Deutsch borrows from philosopher Karl Popper the idea that even a mechanistic universe must harbor surprises. Popper, who believed that the mind and body constitute separate substances, argued that because we cannot anticipate the thoughts and actions of our descendants the future must be an open book. Deutsch preserves Popperian free will by enabling each cognizant being a free choice of paths within the multiverse. Therefore, though the multiverse is immutable, the route one takes within it forms a matter of personal preference.

One feature that distinguishes Deutsch's approach from its antecedents is that it treats moments of history (individual as well as universal) as parallel worlds in their own right, as isolated from each other as maritime kingdoms. Though each island of instantaneous time occupies an archipelago of causality with those said to come before it and those said to follow it, no currents of forward flow carry conscious observers from one islet to the next. The illusion of temporal change, according to Deutsch, stems from our knowledge that past moments appear different from the present. As Deutsch relates:

> We do not experience time flowing, or passing. What we experience are differences between our present perceptions and our present memories of past perceptions. We interpret those differences, correctly, as evidence that the universe changes with time. We also interpret them, incorrectly, as evidence that our consciousness, or the present, or something, moves through time.[1]

If time doesn't flow, then how do we accumulate memories? As the Oxford theoretician describes, we do so in the same manner as a computer accesses the data on its hard drive. The laws of physics allow certain sectors of the multiverse—what we call our experiences—to be accessible to us. By scanning the features of these regions, we imagine that we once inhabited them. This process, as he defines it, bears a striking resemblance to programs that convert collected information into "virtual reality" imagery.

Because every other conscious being in the multiverse, at any given moment, does exactly the same thing, the illusion is never broken. Each is perfectly convinced that he has left a full life behind him and awaits a meaningful destiny ahead of him. Thus, in essence, nostalgia for the past, along with the pursuit of the future, comprise rhapsodies to lands we've never been to and never will visit.

Given his unique portrait of time, it is interesting to examine how Deutsch considers the question of time travel. Because he does not believe in the flow of time, trivially, he rules it out, as he would even the prospects of someone going to sleep today and waking up tomorrow as exactly the same person. However, if one redefines time travel to mean accessing parts of history that one normally wouldn't visit, he considers it eminently possible. Nothing in his theory, he feels, would exclude the possibility of someone being twenty years old in contemporary times and twenty-one years old in medieval France—as long as physics permits such a displacement.

The latter condition is an important caveat. Deutsch realizes that the question of whether or not quantum gravity allows backward-directed time travel is a matter of controversy. If, in developing a full "theory of everything," researchers ultimately prove that time travel is physically impossible, he would exclude it from his own model as well. Deutsch's model encompasses only physically realizable worlds; no more and no less.

If, on the other hand, backward time travel is reasonable, the multiverse approach would help resolve the three serious quandaries we've mentioned: the grandfather paradox, the "something from nothing" paradox, and the "changing one's mind" paradox. In a vast web of parallel universes, in which all scientifically possible combinations occur somewhere, the grandfather paradox would cease to be an issue. Whenever somebody traveled back in time, he would land, not in his own time line, but in an alternative realm. Therefore, whatever heroic acts or misdeeds he committed would affect a different parallel domain from the one in which he was born. It would be absolutely impossible, in that case, for him to make the mistake of negating his own birth.

The "something from nothing" dilemma would resolve itself by means of Deutsch's concept that time has no flow. Because in some sense, every moment in time would maintain an independent existence, the idea of something appearing out of the blue because its cause lies far in the future would no longer seem so shocking. Causality would cease to be a question of continuity, and become an issue of physical allowability. As long as time-reversed patterns of cause and effect were physically realizable, they would present no conceptual problem.

Finally, if somebody willed an action involving backward time travel, then changed his mind, the cause of the widowed effect would be clear. He would not have to carry out the deed himself; an alternate version in a parallel universe could readily complete the task. Hence, as in the case of the other two paradoxes, Deutsch's model provides a means of understanding the situation.

Deutsch's multiverse approach is not for everyone. Considering our intuitive sense of the flow of history and our vision of the continuity of life, it's hard to grapple with the notion that every tick of the clock characterizes a completely autonomous instant. Could everything we know and feel regarding the unity of existence constitute a mere illusion? Nevertheless, if time travel eventually turns out to be workable, Deutsch's model offers possibly the fullest resolution of seemingly intractable dilemmas associated with its operation.

The concepts of destiny advanced by Everett and Deutsch resemble ever-branching mazes, in which each path transports one to a different universe. Yet standard quantum theory is no less labyrinthine. The main difference is that in the Copenhagen approach, fate's labyrinth completely resides within our own cosmos and constitutes a dynamic observer-dependent entity. In other words, in the conventional approach, one cannot pull the magician's trick of positing secret trapdoors into other worlds—its multitudinous mysteries evolve right before one's eyes.

Reality's Lexicon

Richard Feynman was one of the supreme advocates, popularizers, and enhancers of quantum theory—as described by its standard interpretation. By providing a remarkable scheme—known as Feynman diagrams—for summing up the interactions between particles, he created a language for describing an almost inconceivable realm in which a wide range of contradictory possibilities might take place simultaneously. Without his well-crafted lexicon, it is doubtful that theoretical particle physics would have progressed so far so quickly during the latter half of the twentieth century.

Feynman could well appreciate the puzzling contradictions contained in quantum physics. Working part of his life on secret government programs, including the Manhattan Project, he enjoyed the challenge of cracking open safes holding classified documents. A New Yorker by birth, he took to the culture of southern California (where he was a professor at Caltech) with great alacrity. Friendly and open to

FIGURE 4.2   *Richard Feynman (center), Stanislaw Ulam (left), and John Von Neumann (right), three pioneers of prediction (courtesy of the AIP Emilio Segre Visual Archives, Ulam Collection).*

people of sharply contrasting backgrounds, he felt equally at home discussing the fine points of physical theory with Einstein and Bohr and chatting about gadgets with workmen. His bongo playing formed perhaps as important part of his self-image, as did his talks at conferences. Shortly after winning the Nobel Prize in Physics, a feat about which he felt some ambivalence, he took great pride substituting for Italian actress Gina Lollobrigida as the official guest of honor reviewing Samba parades during the Carnaval celebrations in Brazil. Though he was an esteemed professor, thumbing his nose at stuffy tradition gave him enormous pleasure. In a famous incident, when asked at a formal afternoon tea in Princeton if he preferred cream or lemon in his tea, he replied "I'll have both, thank you."[2]

Considering the rich palette of contrasts exhibited in his personality, no wonder Feynman could address so splendidly the puzzles and paradoxes of quantum theory. In developing the field with which he is most identified, quantum electrodynamics, he helped to elucidate an

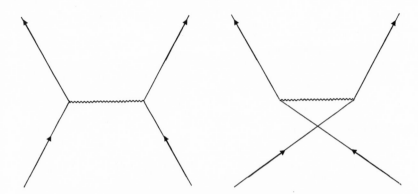

FIGURE 4.3   *Shown are two Feynman diagrams, each depicting the mutual scattering of two electrons (drawn as straight lines) by means of a photon (drawn as a wavy line). Although in the diagram on the left-hand side the electrons simply bounce off each other, in the one on the right they exchange places. Because a researcher detecting the scattered particles would not be able to tell which electron is which, reality embodies a composite of the two possibilities. In general, as Richard Feynman demonstrated, the history of any set of particles might simultaneously embrace numerous possible interactions, each representing an alternative pathway through time.*

area that had baffled theorists for decades: namely the process by which electrons and other charged particles interact with photons. (Along with Feynman, two other physicists, Julian Schwinger and Sin-Itiro Tomonaga, cofounded the field, but expressed their results in forms generally considered less accessible.) Feynman's profound analytical abilities and appreciation for expressing solutions to complex issues in the simplest possible manner helped fashion his formulation of this novel approach.

Feynman initially developed his scheme as a kind of personal shorthand. Whenever he calculated interactions between electrons and photons, he found himself doodling what he called "funny pictures." He'd start out by designating "time" as the vertical axis and "space" as the horizontal axis of a graph. For each electron, he'd draw a straight line and label its direction of motion with an arrow. He'd do the same for photons, only with wiggly lines instead of straight. Moreover, to indicate whenever one particle reacted with another, he had their lines intersect.

To evaluate the physical properties of a given set of particles, Feynman drew each of the possible diagrams corresponding to feasible interactions that might take place among them. For example, two electrons, in exchanging a photon, might either bounce off each other or trade places. Feynman accounted for each of these alternatives by composing separate representative figures. Then, by using calculus to add the diagrams, he obtained theoretical values for quantities such as cross section (a measure of the amount of particles deflected at given angles), which corresponded remarkably well to known experimental results.

Feynman's vision of quantum physics offers a complex portrayal of destiny—all the more unsettling than the Many-Worlds model because its drama takes place entirely within the theater of our own universe. Somehow, in every physical interaction, all the alternative scenes are played out at once—with only the final results discernible to onlookers. In other words, in the quantum world, the ends cannot be used to ascertain the means.

Imagine attending a Shakespearean performance with dual casts, in which Caesar kills Brutus on one side of the stage while at the same time Brutus kills Caesar on the other side. Later, in these simultaneously acted alternative renditions, whoever is the murderer takes his own life in despair. By the end of this strange (dadaist, perhaps) version of *Julius Caesar* both principal figures would be dead, and it would be up to the audience to decide in which of the two "realities" they believe. Such is the strange multiplicity of Feynman's theory of electromagnetic interactions.

Quantum electrodynamics represents one of the twentieth century's most successful predictive models. Yet when Feynman unveiled his scheme to Bohr, it was received rather coldly (as was Schwinger's presentation of equivalent results). The venerable Dane detested the young American's casual approach to the sacred Copenhagen interpretation, and saw his squiggly diagrams in the same light that an expert on Monet would view subway graffiti. Even though Feynman explained that he meant them only as bookkeeping devices, Bohr railed on that the uncertainty principle precluded drawing such precise figures. As Feynman later recounted that depressing meeting: "[Bohr] said that this idiot didn't understand quantum mechanics at all. My ideas were consistent. I knew that his objection was wrong." [3]

Strange how one era's innovator might become the next era's conservative. In destiny's game, players' roles are often reshuffled. History seems to enjoy exploring all possible alternatives.

We have seen how relativity and quantum physics have impacted on the science of forecasting. In comparison to their frenzied, labyrinthine constructs, classical Newtonian dynamics appears relatively tame. But appearances can deceive. In recent decades, researchers have revealed that even the simplest, deterministic set of equations may generate solutions of astonishing complexity. Furthermore, discerning how simple principles might sprout wild, complex behavior, they have discovered abundant order buried within these tangled fields of chaos. Tapping into these rich underlying patterns has offered predictive science its subtlest challenge.

5

CHANGE IN THE WEATHER

Chaos and Prediction

O wild West Wind, thou breath of Autumn's being,
Thou, from whose unseen presence the leaves dead
Are driven, like ghosts from an enchanter fleeing.
                                    —PERCY SHELLEY
                                    (Ode to the West Wind)

Back to Earth

Those who ascend to the dizzying heights of relativity and quantum theory comprise the sturdiest of climbers. Nevertheless, after breathing in the thin air of abstract discussions at quantum gravity conferences, even the most stalwart scholars occasionally grow light-headed. Suddenly, getting back to Earth and working on more mundane problems seems like the most sensible course.

From the 1940s to the 1960s, Nobel Prize–winning pioneers such as Richard Feynman, Julian Schwinger, Sin-Itiro Tomonaga, Sheldon Glashow, Steven Weinberg, Abdus Salam, and Murray Gell-Mann constructed dramatic new theories of elementary particles and forces that lent excitement to the field of physics. Feynman, Schwinger, and Tomonaga, in erecting their meticulous edifice of quantum electrodynamics—unprecedented in its accuracy—capped centuries of debate about the nature of electricity, magnetism, and light. Glashow, Weinberg, and Salam built on this lofty construct another exhalted structure, merging fully two of the four known forces of nature—electromagnetism and the "weak force" into their "electroweak model." Then, in a prediction that must be considered one of the highlights of twentieth-century physical forecasting, Murray Gell-Mann employed symmetry principles to surmise correctly the existence of quarks—the fundamental constituents of many types of elementary particles.

But what does a creative individual do once the beautiful buildings have been completed and the architects of these creations have collected their accolades? Settle for more mundane projects, or try to build higher and higher? In the 1970s, many opted for the latter, and attempted to construct even more glorious theories, encompassing not just electromagnetic and weak forces, but also the strong force (that glues together quarks), and gravity. Building on the electroweak approach, physicists such as Howard Georgi of MIT suggested profound models of unification. But sadly, none of these mathematical creations were found to correspond to physical reality. Field theory, it seemed at the time, could ascend no further.

Not only were theorists caught up in a tangle of increasingly treacherous mathematics, their models went increasingly untested. The existing particle accelerators, large enough to confirm the hypotheses of the 1960s theoreticians (the quark and electroweak models), fell short of testing anything more comprehensive. The cost of high-energy physics soared to depressing new heights as the energies required for new experiments skyrocketed. Particle experimentalists found it increasingly difficult to justify their requests for larger and larger shares of the research pie. As government budgets tightened, it looked like only a handful of new colliders could be supported. Theorists wondered what they would do once these were built and they needed even more power to test their hypotheses. (Even that view turned out to be too optimistic. The major American project of the 1980s and 1990s—the superconducting supercollider scheduled to be built in Texas—was canceled in midconstruction because of cutbacks.)

Given these depressing statistics, universities began to emphasize cheaper, more accessible, "table top" experiments. Those who once dreamed about atom-smashing began to consider simpler projects instead—analyzing the turbulent behavior of water waves or measuring

FIGURE 5.1 *Murray Gell-Mann (1929–), Nobel Prize–winning developer of the quark model, has more recently lent his efforts in trying to resolve fundamental questions in quantum and complexity theories (courtesy of the Santa Fe Institute; photograph by Murrae Haynes).*

the erratic effects of certain types of pendula. Acoustics, hydrodynamics, meteorology, geophysics, and other aspects of earth science increasingly attracted brighter and more eager students.

Another tendency of the times was steering young researchers in this direction—the "back to Earth" movement in general. In reaction to the Vietnam War, the Cold War, and the nuclear age in general, many sought solace in simple lifestyles. New communities devoted to living in harmony with nature attracted thousands of enthusiastic members. Naturally, a large portion of those with scientific inclinations and who felt politically attuned to ecological concerns considered combining their interests in earth science–related careers.

Many of these young scientists soon realized, much to their surprise, that a large number of natural systems—although grounded in fundamental Newtonian mechanics—displayed astonishingly complex behavior. Though in principle such phenomena were driven by deterministic rules, in practice they acted in highly unpredictable ways. Young researchers had fun testing the seemingly random nature of these simple mechanisms. The lure of reexamining basic science through a fresh perspective and discovering something new about the world in the process proved enticing indeed.

The phenomenon of a simple deterministic system behaving in an apparently sporadic fashion soon acquired a catchy name: "chaos." For those motivated by an inner compulsion to understand the underlying mechanics of nature, chaotic dynamics, self-organization, fractals, and related subjects offered exciting new ways of interpreting reality. But did chaos theory mean that prediction was impossible? Or was there a "hidden order" that would ultimately reveal aspects of the future?

Rambling, Gambling, and Equation Scrambling

James Doyne Farmer, who calls himself by the second of his given names (pronounced to rhyme with Owen), came of age in the freest era for self-expression since the days after World War I. In the same manner that intellectuals of the late 1910s and early 1920s drank a rousing brew of dadaism, jazz music, relativity, and quantum physics, he, along with the rest of his generation, quaffed a stimulating concoction of radical street theater, psychedelic rock, communal living, and, ultimately, chaotic behavior in science. Now he is one of the world's foremost experts in forecasting—but who could have predicted that?

Born in 1952, young Doyne grew up in the desert town of Silver City, New Mexico. As a bright, bookish lad in a region known mainly for mining, he found little in common with his fellow students. Often lonely, his isolation was broken through a chance encounter with a physics instructor who was helping out his Boy Scout troop. The instructor became his mentor, even lending him a spare room when the rest of his family temporarily moved to Peru. Through the tutelage of his role model, Farmer became fascinated with the physical sciences—

FIGURE 5.2 James Doyne Farmer (1952–), member of the Dynamical Systems Collective and cofounder of The Prediction Company (courtesy of the Santa Fe Institute; photograph by Murrae Haynes).

particularly the study of the stars and planets. Common interests brought him another connection—friendship with Norman Packard, another member of the troop, who ended up becoming a long-term scientific collaborator. Finally, Farmer had found people with whom he could talk about meaningful ideas.

The seventies began in a tumult of protest. As a vocal pacifist, during an era of what many saw as senseless warfare, Farmer identified himself with the youth movement of the day. He grew his sandy hair long, became an avid biker, and set out on his motorcycle (after a short time in Idaho) for the West Coast center of action: the San Francisco Bay Area. He studied physics as an undergraduate at Stanford, where he lived in communal quarters with other hippies, and played blues harp in a local rock band. After obtaining his degree, he went to the · University of California at Santa Cruz, where he began a course of study in astrophysics and relativity. Packard, who was then a student at Reed College in Portland, kept in touch and visited often.

During one summer at Santa Cruz, Farmer took a break from his graduate work on galaxy formation, and, while working a stint at the U.S. Forest Service, took up an interest in poker playing. He later referred to this interlude as the start of his "rambling and gambling."[1] Packard, in the meanwhile, had discovered Las Vegas, and spent his days in the casinos engaged in working the blackjack tables and trying his hand at the roulette wheel. After the summer was over, the friends met up to compare notes.

Many experts in the predictive sciences become attracted, at some point in their lives, to gambling. Those whose inner fire compels them to understand how best to anticipate the future, often cannot resist expressing their talents in a series of well-played hands. The casinos are naturally well acquainted with statistical ways of beating the dealer. Typically, they tailor the odds to their advantage, and ban all methods, such as card counting, that might be used to tip the tables. Therefore, even the best brains find it hard to win consistently in casino-organized card games.

Normally the odds for roulette, from a player's point of view, are even worse. However, unlike poker and blackjack, everything that happens to a spinning wheel is out in the open. Theoretically, if one fully understood the dynamics of a roulette wheel, knew its initial state, and estimated the velocity given to it by the spinner, one could predict where it would land. Thus, in principle, a clever observer with access to a sufficiently precise algorithm and a fast enough means of calculation could beat the odds and take home money from the game.

Packard pointed these facts out to Farmer. At first he was skeptical, but then, after some estimates and some experimentation, he became convinced that they could make a nice profit. Along with some of their other friends, they formed a group to carry out their casino-beating plan code-named Project Rosetta Stone, or the Project, for short. In the collective spirit, they would contribute equally to the Project, then equally divide whatever profits they made. The group worked hard—testing out the properties of roulette wheels—and determined that they could carry out their goal if they could fashion a small enough computer to bring with them to the casinos.

In the 1940s and 1950s, when computers were first developed, their clumsy components—rigged up out of metal cases, vacuum tubes, and tangled wires—stretched out across good-size rooms. To increase their computational power, scientists and science fiction writers alike (including Isaac Asimov in his famous story "The Last Question") imagined that they would be constructed bigger and bigger. In one of the most-glaring failures of modern prediction, no one anticipated that computer central processing units would shrink down to miniscule proportions, painted onto silicon chips.

By the time of the Project, in the mid- to late 1970s, Farmer, Packard, and their collaborators realized that they could construct computational devices small enough to fit into the heels of their shoes. They did just that, fashioning devices that they could operate with their toes. Tiny transmitters radioed their data from one operator to another as they spread out across casino floors, collecting and processing roulette wheel data. The routines programmed into their minicomputers used Newtonian mechanics to forecast in which regions the wheels would land. This information would be relayed to the Project members chosen to place the bets. Over time, using this method, the group made a reasonable amount of money—not a fortune, but enough to buy new equipment and keep on going. Any profits left over were dutifully divided among all participants.

Farmer commuted back and forth between Nevada and Santa Cruz, where he maintained his graduate student status. Packard, in the meanwhile, had begun his own studies at Santa Cruz as well. With astrophysics and standard statistical mechanics (Packard's field) seeming too remote and ethereal, they looked for more compelling projects with which to become involved. The study of chaotic dynamics, related as it was to their interest in prediction, fit the bill perfectly. Together with two other Santa Cruz students—Boston native Robert Shaw and Californian James Crutchfield—they formed another research group, this one more academic in emphasis. They called it the Dynamical Systems Collective, but it quickly became known as the Chaos Cabal. Its goal was to unravel the behavior of strange attractors and other curious objects associated with the rising new field founded by Edward Lorenz.

Wing Flaps and Thunderclaps

Edward Lorenz, the universally recognized guru of the chaos movement, stands as an unlikely revolutionary. A friendly, soft-spoken meteorologist, he is more likely to blend in with the crowd than to summon its attention. Though his weathered expression betrays his occupation, his luminous eyes form the only aspects of his features that give away the radical impact of his work.

Before Lorenz, Laplacean determinism held unchallenged dominion over classical physics. Sure, relativity, to some extent, and, quantum physics, even more so, raised doubts about how to interpret the clockwork mechanisms of nature. But each applied mainly to exotic domains—the very small, the very large, and the very fast—and neither conflicted with deterministic ideas on a mundane level. Poincaré had brilliantly pointed out the limitations of Laplace's work, but for decades no one seemed to follow up on his theories.

Lorenz himself started out believing wholeheartedly Laplace's dictum: given the essential initial conditions of a problem and the proper dynamical equations, its future behavior might be entirely forecast. In 1960, he developed a simple computational model of the weather—a set of equations relating wind velocities, temperatures, pressures, humidity, and other meteorological phenomena—that he fully expected to operate like clockwork. Predicting the weather, he knew, involved applying Newton's mechanical laws to the atmosphere; nothing more, nothing less. With the power of a computer to carry out such calculations, he believed, forecasting wind, sleet, snow, and hail would soon become a matter of routine.

In fact, when the electronic computer was first designed by mathematician John von Neumann and others, one of its two intended primary purposes was enhanced weather prediction, the other being military use. Since the days when farmers turned to almanacs for advice on when to anticipate cold snaps, heat waves, dry spells, and rainy seasons, predicting the weather well ahead of time has been a long-held dream. But for centuries no one could do much better than to offer aphorisms such as "red sky at night, sailor's delight; red sky in the morning, sailor take warning."

In 1904, modern meteorology began in earnest when Norwegian scientist Vilhelm Bjerknes divided Earth into a grid pattern and noted the atmospheric conditions in each sector. Several years later, British mathematician Lewis Richardson set a record by forecasting the weather fully *six hours* in advance. By 1953, led by von Neumann, the Institute for Advanced Study Meteorology Project was able to make the first twenty-four-hour forecast, using only six minutes of computer time. Naively, many thought that with better computers, accurate meteorological forecasts weeks or even months in advance would soon be possible. In designing his simple model, Lorenz was hoping to take the first steps toward such a goal.

When Lorenz tested his crude program, he did not expect much. He had made too many simplifications to hope for anything close to actual forecasts. Nevertheless, he was curious to see what would happen. He plugged some numbers into his algorithm, and out came a weather forecast, which he printed out as a graph. With its rugged peaks and valleys, it looked splendid, matching his intuitive understanding of meteorological phenomena. Very good so far, he thought.

To examine his results closer, he started up his program again, typing in what he believed were the same numbers. Strangely, the printout looked completely different; crests existed where there formerly were troughs, and vice versa. He double-checked the values that he had entered and discovered that he had rounded them off, dropping a few of the final digits. Because the numerical difference between the rounded and original values was miniscule, he was astounded that his minor change caused such a huge discrepancy. Moreover, he tried this substitution out with other sets of data, and found similar results.

Lorenz had discovered the "butterfly effect": the phenomenon that, for certain deterministic systems, a small discrepancy between one set of initial values and another might magnify very quickly over time into major differences in resulting behavior. Therefore, a purely mechanistic series of equations might in the long run yield wholly unpredictable results—seemingly as random as a coin toss. Because nature harbors many built-in variations—courtesy of the uncertainty principle, as well as of statistical fluctuations in general—such imprecision would be common. Perhaps, then, the turbulent behavior exhibited by many physical systems might be explained by such mathematical oddities.

Lorenz deemed his discovery the "butterfly effect" because of its implications for the weather; even a small creature's movements might generate an enormous influence. Theoretically, for example, the flapping of a butterfly's wings over a baseball field in Sioux City, Iowa, might cause a thunderstorm during a soccer match in Nuku'alofa, Tonga, on the other side of the globe. Because of the amplification built into the dynamics of the weather, the butterfly's minute influence

FIGURE 5.3   The devastating power of a tornado exemplifies the onset of turbulent, unpredictable behavior in a deterministic system such as the weather (courtesy of the National Oceanic and Atmospheric Administration).

would grow over time to become a whopping meteorological disturbance.

When Lorenz published his initial results in a 1963 paper titled "Deterministic Nonperiodic Flow," it went, at first, virtually unread. Unfortunately, few mainstream physicists perused atmospheric science journals, and few atmospheric scientists speculated about universal implications. Like a long lull before a swift downpour, his findings created little stir until the mid-1970s, when suddenly the interpretations of his work by other scientists generated a torrent of interest.

Some Strange Attraction

At the end of his paper, Lorenz sketched a mysterious object—a mapping of his weather data in three-dimensional space. He successively plotted the values of three of the variables describing meteorological conditions, and showed how they outlined a ghostly object—shaped just like a butterfly, strangely enough, with its two winglike portions. He indicated how any two nearby points, chosen on one of the wings, would dynamically evolve in such a way that their descendants would end up on opposite sides. Moreover, any points selected off the wings would develop into values on one of the wings. In short, points on the object would spread out over time; points off of it would converge over time. It's just like migration to the suburbs; many are attracted there, but, once there, want to move as far away from each other as possible.

In 1971, Belgian physicist David Ruelle and Dutch mathematician Floris Takens, in a theoretical paper on turbulence, came up with a name for such objects: strange attractors. They were looking at such figures in the abstract, investigating their propensity to fold up space in some directions and stretch it out in others. Ordinary attractors in a data set tended, like flypaper, to collect points. These apparently did something more—mixing up what they collected.

Other examples of strange attractors were soon found, both in theory and practice. A stretching and folding algorithm developed by French astronomer Michel Hénon was shown to yield such a piece. As in the case of the Lorenz attractor, points in the Hénon attractor jumped around in a curious cacophony, while honing in on thin ultimate structures. American experimentalists Harry Swinney and Jerry Gollub pioneered investigating strange attractors in table-top experiments measuring turbulent fluid flow. Soon, no subject was off limits for investigation—strange attractors were found in everything from heart rhythms to Saturn's rings.

The ubiquitous nature of these curious objects was perhaps best demonstrated by Robert Shaw, one of the founders of the Dynamical Systems Collective. He chose one of the simplest possible systems: a slowly leaking faucet. With a stopwatch, he recorded the times between successive drips of water falling from a tap. Then, taking a piece of graph paper, he plotted the interval between one pair of drips versus the interval between the next pair, again and again, until he exhausted all the data. The points, at first seemingly random, soon fell neatly into a strange attractor pattern. Apparently, not even superintendents of old apartment buildings might escape the clamor of orderly chaos.

Those beholding the splendor of strange attractors could not help but notice their wispiness and their self-similarity. Rather than standing solid, like dense cubes or filled-in disks, these objects appeared to speckle space with fine dust. No matter how closely one looked, gaps opened between neighboring points. Moreover, on every level, the general appearance of a strange attractor seemed the same. Views magnified 10 times, 100 times, and 1,000 times all appeared similar—like images reflected in a hall of mirrors. Unmistakable signatures of what came to be called chaotic dynamics, all strange attractors possessed these distinct qualities.

As if through serendipity, the mathematics of self-similarity flowered in the 1970s around the same time that the science of chaos sprouted. In 1975, French mathematician Benoit Mandelbrot, while investigating questions in geometrical measurement, invented the term fractal to characterize structures similar on all levels of observation. He argued that such entities possessed fractional dimensionality: spatial dimensions other than the usual one, two, or three.

One of his favorite examples of a fractal stemmed from attempting to measure the coast of Britain. Naively, one would think that the length of the British coastline would have a definite value. But Mandelbrot demonstrated that this would not be the case; the result would depend on the scale of the measuring instrument. For instance, delineating the distance with 100-mile rods would skip over aquatic features such as the mouth of the Thames, the Wash, the Firth of Forth, and the Bristol Channel. A 10-mile measuring stick would detect these large inlets, but miss smaller mouths of rivers and streams. Mile-sticks and yardsticks would each pick up on these, but ignore yet other features—such as rocks. As finer and finer instruments were used, greater and greater coastline lengths would be obtained.

The reason for these discrepancies, Mandelbrot showed, is that the British coastline has fractal features. It is roughly self-similar on all levels—from the dimensions of broad channels down to the size of tiny pebbles. Its dimensionality, he demonstrated, lies between one and two—not quite a regular curve, but not quite a filled-in shape either.

Fractal dimensionalities might be computed for various strange attractors. Typically, if a strange attractor can be embedded (placed) in a plane, then it has a fractional dimension somewhere between one and two. However, if it spreads out throughout a three-dimensional space, not quite filling it, then its dimension generally lies between two and

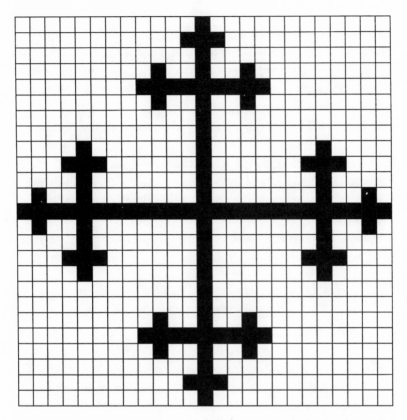

FIGURE 5.4 Here is a basic grid pattern that, which may be used to generate a cellular automaton. The black squares in the pattern correspond to the value "1" and the white squares to "0." Applying simple rules to this initial state (those of Conway's Life, for example), one might discern the future values for each cell. Note that the black squares in this pattern are arranged in the self-similar mathematical form known as a fractal.

three. The higher the fractal dimensionality, relative to the space in which it is embedded, the more intricate the pattern.

In the late 1970s the Santa Cruz Dynamical Systems Collective launched a full study of chaotic behavior, seeking mathematical ways of characterizing the properties of strange attractors. Robert Shaw, who had become familiar with the research of Lorenz and others, served as the catalyst for the group's high-powered activity. James Crutchfield, a brilliant young computer whiz, started out as the team "hacker," then soon came into his own as a major researcher. Norman Packard, who had passed all of his first-year exams in record time, served as the expert in statistical methods. Finally, Doyne Farmer, who realized that he had lost interest in astrophysics and couldn't base his dissertation on the Roulette Project, took up chaos as a new mantra and contributed his own energy and vision to the group.

Farmer recalls some of the reasons that led him to switch fields:

There was a feeling that it wasn't clear where these fields [astrophysics and gravitational theory] were going to go. Would there be innovation or would they be stuck? The kinds of questions addressed in astrophysics and field theory seemed remote. [In chaos theory] there was really new and exciting stuff happening—fertile for innovation and discovery.[2]

Farmer considers himself lucky that Shaw introduced him to chaos theory at that critical time in his life. In the acknowledgements of his dissertation, "Order within Chaos," he wrote:

Had Rob never heard of the Lorenz attractor none of this would have occurred. I would have likely have gotten bored with physics and dropped out, and would now be happily playing my harmonica for the hippies down on the mall. Instead, Rob planted the seeds of chaos in my brain, and here I am trying to be a respectable scientist. Such is sensitive dependence on initial conditions.[3]

For a number of years, the Chaos Cabal brought its energies to bear on critical new questions in dynamics. How might one distinguish between purely random behavior, created by arbitrary disturbances, versus deterministic chaos, generated by mechanistic Newtonian systems? What types of measures characterize the order inherent in chaos? And, finally, what does chaos theory tell us about the nature of free will versus determinism?

The answers to the first two questions they found by examining various dimensionalities and entropies (measures of disorder). Purely random systems, when graphed, tend to fill up the entire multidimensional space. For example, wholly random behavior (rolling dice, for instance) plotted on a two-dimensional graph yields an image that is roughly two-dimensional. The actions of deterministic chaos, on the other hand, visually comes across as lower-dimensional plots.

Attempting to resolve the question of free will versus determinism represented, perhaps, the most ambitious of their projects. They viewed strange attractors as opportunities provided by nature for spontaneous change. Though theoretically these sets might be described by simple, completely deterministic sets of equations, in practice, their dynamics might never be predicted. To make such a forecast, one would have to supply a computer with an entire set of initial data, *precise to an infinite number of decimal places.* Obviously, this would be unattainable. Hence chaos offers the prospect of changes impossible to anticipate, creating a natural opening for the actions of free will. Laplace's demon loses his prophetic powers.

Another way of stating this is that chaotic systems are *computationally irreducible.* The state of the system, at any given time, might only be "anticipated" by carrying out all of the dynamics step by step. Therefore, a chaotic system is its own fastest computer. Like a good detective story, to figure out the ending, one must read carefully every single page; no shortcuts allowed!

The implications of these findings for human destiny are staggering. If nature is chaotic at its core, it might be fully deterministic, yet wholly unpredictable. The only way to anticipate the future, then, would be to run through all of history from now until then, step by step. But the fastest way to do this would be to wait for history itself to unfold. Hence even a clockwork universe might harbor its share of surprises.

The Limits of Computation

In the modern technological age, a commonly held assumption is that computers might in theory perform any "mental" task, given enough time and resources. According to this myth, as computational power increases, all conceivable scientific problems will find solution through machine-based algorithms. Yet a number of advances in logic, some dating back to the 1930s but not fully grappled with until recent decades, have called into question this line of thinking.

The most fundamental of these, Gödel's Incompleteness Theorem, was published by Austrian mathematician Kurt Gödel in 1931. One of the greatest mathematical achievements of the twentieth century, it was proposed as a rebuttal to the Formalist view put forth earlier by mathematician David Hilbert. Hilbert had attempted to show that all of mathematics might be represented in a manner that is logical, consistent, and complete. Gödel's Theorem demonstrated that no such system exists; in fact, according to the laws of logic, it cannot exist.

Pythagoras said that "number is all." Kepler concurred, for a time, placing geometrical considerations such as symmetry and regularity above empirical evidence, until facts forced him to conclude otherwise. The early quantum theorists thought, in some sense, that "wavefunction is all." Philosophically their work marked a revival (in spirit, not substance) of Plato's notion of Forms. Now, those championing the decoherent histories approach argue for a return to an observable-based dynamics. The question of whether or not abstract quantities might embody the universe in and of themselves without recourse to connections with the physical has been debated for millennia—with a wide range of philosophers arguing the Idealist (abstract) and Realist (concrete) positions. And, until Gödel addressed Hilbert's Formalist conjecture, neither camp had furnished definitive proof of its position.

Hilbert's Formalist program, launched in the early decades of the twentieth century, was an attempt to frame all the axioms and theorems of mathematics in purely symbolic form, decoupled from their references to actual objects. He recognized that each mathematical term, such as "circle," plays two distinct roles: semantic and syntactic. A term's semantic function involves a mapping between its symbolism and its physical form; for example, the abstraction spelled "c-i-r-c-l-e" represents a perfectly round geometric figure. Its syntactic role, in contrast, involves the way it is manipulated with other terms to produce logical statements.

In some cases, syntax alone might make perfect sense. If a girkle is a smirkle, and a smirkle is a tirkle, then logic forces us to conclude that a girkle is a tirkle, even if we don't know what these terms mean. Or, if someone says: "Dr. X had lunch with Mr. Y," we naturally deduce that Mr. Y had lunch with Dr. X as well.

Hilbert tried to do this for all of mathematics—beginning with a theory of arithmetic—attempting to fashion an entire system, fully consistent, out of pure symbolic reasoning. If such a feat could be done, no one doubted that he had the credentials to do so. Along with Poincaré, Hilbert was considered the greatest mathematical genius of his day. Yet, as Gödel showed, not even the most ingenious mathematician of all times could complete such a program.

Gödel turned Hilbert's work on its head by developing an "arithmetic of logic," and then showing that any logical formulation of arithmetic must address the "arithmetic of logic" as well. First he assigned a number to each symbol in logic: "1" to "not," "2" to "or, " and so on. Then he demonstrated that any logical statement (Either A is true or A is not true) might be encoded in terms of mathematical functions of these numbers. Finally, to clog Hilbert's carefully constructed machinery, he asserted that the following expression, translated into its numerical equivalent, must be included in any formalism of arithmetic: "This statement is not provable."

Note the marked similarly between this expression and the Epimenides paradox, "this sentence is false." In either case, the statement has no defined truth value. Because its validity is indeterminate, it must lie outside any self-consistent logical system that attempts to assign truth or falsity to any conjecture. Yet Gödel showed that *any* complete theory of arithmetic must include it. Consequently, one cannot construct a self-contained, entirely consistent theory of arithmetic.

Gödel's Theorem represented as crushing a blow for modern mathematicians as the existence of irrational numbers was for the Pythagoreans and quantum randomness was for Einstein. It showed that how no matter how much of mathematics is codified, there will always be gray areas. As mathematician John Casti has remarked on this point, "There's no washing away the gray!"[4]

Gödel's Theorem is clearly one of the greatest conceptual advances in twentieth-century predictive science. Along with the uncertainty principle and chaos theory, Gödel's Theorem strongly suggests that, in the words of Farmer, "There are always going to be inherently unpredictable aspects of the future."[5]

There is another problem that illustrates the fundamental limits to knowledge. A conundrum in computer science integrally related to Gödel's Theorem, the Turing Halting Problem was discovered by Alan Turing as a consequence of his instrumental work in the realm of artificial intelligence. It concerns predicting whether a computational procedure can be completed in a finite amount of time. Turing demonstrated that in many cases such questions are indeterminable; that is, no one might know in advance if they will finish.

To understand the Halting Problem, we must first address Turing's concept of an abstract computational engine, called a Turing machine. A Turing machine is a hypothetical mechanism designed to perform an indefinite range of algorithmic tasks—from adding two numbers together to computing the tenth digit of pi. It consists of two elements, designed to work in tandem. The first component is an infinitely long tape, marked off into squares like a roll of film. Each square contains a symbol, selected from a finite set. For example, if the set consists of zeroes and ones, the first square might contain "1"; the second, "0"; the third, "1" again, and so forth.

The second part of a Turing machine comprises a scanning device with an internal gauge that might be in one of a finite set of states, called "alpha," "beta," and "gamma," for instance. Think of the gauge as a pointing arrow, aimed at one of the possibilities—"beta," let's say. The scanner reads the tape, one square at a time, and performs certain tasks depending on which internal state it is in and which symbol it detects. The actions it might perform include the following: moving the tape one square to the left, moving it one square to the right, and stopping the entire process. At the same time it either writes a new symbol onto the tape or preserves the old symbol. Finally, it either keeps the same internal state or sets the gauge to a new state. For example, if the symbol read on a certain square is "1" and the internal state is "beta," the machine might be programmed to change the "1" to a "0," advance the tape one square to the right, and reset its gauge to "alpha." The device keeps going until the stop command is executed.

Though much has been written about Turing machines, two critical theorems related to them seem particularly pertinent to this discussion. The first, called the Church-Turing thesis, or sometimes Church's thesis, was independently proposed by Turing and the American mathematician Alonzo Church. It states, in essence, that any effective method might be executed on a universal Turing machine.

An effective method, as defined by Church, is a mechanistic procedure that can be carried out in a finite number of steps with no particular insight or ingenuity needed. An example of this is the algorithm used to calculate the square root of a whole number. One might think of countless others—practically any simple arithmetical procedure.

Turing demonstrated that he could construct an abstract computational engine to complete any such effective task. Conversely, if he could not design a Turing machine to solve a proposed problem, it could not be solved by a step-by-step procedure. Its solution would therefore be computationally indeterminate.

The Church-Turing thesis has sometimes been used by workers in the field of artificial intelligence to support the claim that any task the human brain could perform, a sufficiently powerful computer could likewise be programmed to complete. However, this is fallacious reasoning. Undoubtedly, many of the methods used by the brain to tackle a problem—insights, hunches, good guesses, and so forth, could not be construed as step-by step procedures. For instance, a chess master might win a game through a purely intuitive strategy. In such cases, the Church-Turing thesis would clearly not apply.

Nevertheless, for the vast array of problems potentially solvable by algorithm, the Church-Turing thesis poses a quandary. It challenges mathematicians to ponder how Turing machines might resolve them. If no theoretical engine could be constructed to do the job, then similarly no real computer could handle them.

Here is where the Halting Problem comes in. Suppose a math problem is placed on your desk. You want to figure out if it can be resolved by computer. Therefore, you think for a while, and then construct a Turing machine that seems up to the task. You work out a few iterations (steps) of the program, and realize that it will take a very long time for it to find a solution—if it can at all. Consequently, you attempt to establish if the routine will ever reach a conclusion. But, alas, no matter how hard you try, you cannot ascertain if the program will take a finite—or infinite—number of steps.

Like Gödel's theorem, the Turing Halting Problem places an impenetrable curtain between the realms of what mathematicians can and cannot know. As Turing pointed out, there are many computational problems for which algorithms can be constructed, but for which nobody can determine in advance if they have solutions achievable in a finite amount of time. Mathematicians, in such situations, are placed in the position of flustered millionaires, eager to disperse their fortune if they'll see a return, but who do not know whom they can trust to lend their money. With unpredictable outcome and unclear payback, they just have to take their chances.

A more recent example delineating the limits of mathematical knowledge was developed in 1987 by Gregory Chaitlin of the IBM Research Laboratories. Chaitlin, who has been cracking computational quandaries since he was in high school, developed a theorem showing that arithmetic itself—a seeming bedrock of determinism—has qualities as random as a coin toss. In his words: "God not only plays dice in quantum mechanics, but even with the whole numbers."[6]

Chaitlin's theorem concerns a class of algebraic relationships, familiar even to the classical Greeks, known as the Diophantine equations. The form of these equations is determined by a set of parameters, selected from the whole numbers. Depending upon the values chosen for these parameters, the equations may or may not have solutions. Chaitlin varied one of the parameters, called k, and took note, for particular values of that variable, whether a computer would halt in attempting to find a solution to the corresponding equation. He recorded his results as the binary (base 2) digits of a special number, which he called Ω. As he defined it, the kth digit of Ω is 1, if the program does halt for that value of k, and 0 if it doesn't halt. Consequently, Ω is an unending sequence of 0's and 1's.

The curious thing about Ω is its sheer randomness. As Chaitlin discovered, 0's and 1's are sprinkled throughout Ω as evenly and haphazardly as yolk and white throughout a plate of scrambled eggs. No one might ever compute the full number's exact value. Yet the Diophantine equations represent well-defined problems, emblematic of simple arithmetic procedures. Apparently, the specter of chaos haunts even the austere halls of Pythagorean numerical purity.

Chaitlin has pointed out that Ω possesses additional significance. Converted into a decimal number for convenience, with a decimal point placed in front of it, it constitutes the probability that a randomly chosen Turing machine program will halt. (Ω can't be computed exactly, but it can be approximated.) It offers a concrete measure for delineating how much complexity and indeterminism there is in mathematics. Thus, though Chaitlin has unveiled strata of chaos beneath the surface of arithmetical truth, he has found in Ω an excellent "geological" measure of how much of this buried disorder exists.

The Path to Chaos

Revealing order in chaos by defining appropriate quantitative measures was an emblem of dynamical systems theory throughout the late 1970s and 1980s. As scientists from all disciplines realized the pervasive nature of apparent randomness, they scanned these turbulent seas for island patterns and were heartened to find them everywhere. From fractal dimensionalities to universal constants (such as Ω), they discovered ubiquitous regularity in certain aspects of chaotic systems.

Biologist Robert May grew up in one island nation—Australia—and ended up in another—the United Kingdom—where he is now chief science adviser to the government. In between, he studied, among other ecological systems, the behavior of fish populations, and chose to do so when he was not on an island at all; rather, he was in Princeton, New Jersey.

Like most groups of creatures, fish populations, in the absence of predators, tend to grow until they have saturated their environment. Once they have achieved the numbers to exhaust the food supplies of their habitats—ponds, streams, rivers, and so on—they start to die out. Typically, enough fish expire until there is plenty of food to go around once again. With greater resources, the population restocks, until it once more bumps up against the fixed limits to growth.

One might characterize the size of such a population, over time, by a relationship known as a logistic equation. The simple form May examined contained one free parameter, called λ, that controlled the population's rate of growth. Like adjusting the throttle of a plane, by cranking up λ, one made the system run faster.

In running his simulation, May noticed a variety of interesting outcomes. For small values of λ, he found that the fish population tended to die out completely. It simply didn't grow fast enough to regenerate itself. Turning the λ dial a little higher, May found a qualitative change in behavior—a sudden transition technically known as a bifurcation. Instead of converging on the value zero, the population grew to a finite, stable limit. Once it reached that upper value, it remained constant forever.

Setting λ higher and higher, May discovered more and more bifurcations. For a certain range of that parameter, the populations would oscillate between two different values—a flipping from one quantity to another called a limit cycle. For greater values of λ, the cycle of two would split into a cycle of four; then, for higher still, a value of eight.

Curiously, though, for some ranges of λ, May found no pattern at all—pure chaos. The system would behave like a random number generator. Yet, emphatically, no external force caused the seeming randomness; like Lorenz's weather model, the chaos swelled up from the deterministic mechanics of the system itself.

Turning up λ even higher, May found a return to periodic behavior. Only this time, the cycles had periods in multiples of three—3, 6, 12, and so on—instead of multiples of two. This would go on for a while, then, for greater λ, a return to chaos.

James Yorke, a mathematician at the University of Maryland and coiner of the term "chaos" to describe seemingly haphazard deterministic behavior, set out to make sense of such patterns of bifurcation. In a startlingly original paper titled "Period Three Implies Chaos," he found the Holy Grail of deciding when chaos will show up. He proved, in mathematically rigorous fashion, that any system churning through cycles of period three for certain values of a parameter must engage in chaotic behavior for other values. Thus, in essence, he discovered a way of predicting when something will appear unpredictable.

Mitchell Feigenbaum, a physicist at Los Alamos, took Yorke's finding a step further. Given free rein to work on problems of his own choosing, Feigenbaum found himself tinkering more and more with simple equations—such as the logistic relationship analyzed by May. He treated the equation as an endless feedback loop, first plugging in a number, obtaining a result, and then inserting that value back into the equation. As May had also seen, by cranking up the equation's driving parameter, he obtained qualitatively different results.

Like a composer delineating breaks between various movements, Feigenbaum noted all the points in which transitions took place. He recorded the value of the parameter for which static behavior (1,1,1,1 . . .) bifurcated into oscillations of period 2 (1,2,1,2,1,2 . . .), then that in which period 2 changed into period 4 (1,2,3,4,1,2,3,4 . . .), and so forth. He noted that these period-doublings occurred more and more frequently with higher values of the parameter. Then, as if the script suddenly changed from classical

FIGURE 5.5 Mitchell Feigen-baum (1945–) (courtesy of Rocke-feller University; photograph by Robert Reichert).

marching music to free-rein jazz, once the parameter rose above a certain threshold, chaos ensued.

Feigenbaum noticed that like the frequencies of notes separated by octaves, the transition values of the parameter seemed to possess similar ratios. That is, the ratio between the onset parameters for period-doubling from period 1 to period 2 was approximately the same as that between period 2 and period 4, and so on. Out of curiosity, Feigenbaum calculated the limit of these ratios for greater and greater values of the parameter, and found that they converged to a particular number, which he computed to three decimal places: 4.669.

On a whim Feigenbaum decided to try the same thing for other equations. He picked a trigonometric function, and began the same iterative process. Once again, he found the same sequence of ratios for period-doubling transitions, and once again he found the same convergence to the mysterious value 4.669. No matter which equations he worked with, his results were identical. Remarkably, he had discovered a new universal property, with its own fundamental constant—akin to pi or "e" (the significant number 2.718 . . .).

Other chaos theorists recognized that Feigenbaum had made a great leap forward. He had shown that inherent order exists, not just in chaos, but also in the transition to chaos. His discovery of a universal constant for period-doubling transitions suggested new avenues for predictive science. Perhaps, they wondered, this curiously omnipresent value could be incorporated into models for forecasting the onset of turbulence—such as tornadoes in weather systems. Yet no one, not even Feigenbaum, could find direct application. The field of chaos matured into the more comprehensive domain of complex systems without—so far—making sense or use of Feigenbaum's strange result.

The Way of Santa Fe

In the early 1980s, the science of complexity rapidly grew as a field. Novel ties were forged as old collaborations broke up. Established physicists joined freshman scientists in exciting new projects.

The Santa Cruz Dynamical Systems Collective, having served its purpose to explore and enlarge upon its subject, elected to dissolve itself once its members received their Ph.Ds. Like seeds in the wind, its members scattered across the country. Norman Packard was appointed to the Institute for Advanced Study in Princeton—Einstein's hallowed ground—where he worked with brilliant young British physicist Stephen Wolfram. Like Farmer, he found in Wolfram another renegade with whom he could exchange ideas. Wolfram, in his early twenties at the time, seemed headed for a promising career in theoretical particle physics. But, as in the case of so many others, he made the jump to complex systems. When Wolfram was selected to head a new Center for Complex Systems Research at the University of Illinois at Urbana, Packard joined him. Individually and in collaboration, they engaged in a meticulous exploration of the collective properties of interacting arrays of elements—systems known as cellular automata.

Doyne Farmer joined the Center for Nonlinear Studies at Los Alamos, a unit founded to foster the investigation of nonlinear dynamics in systems. At the time, Mitchell Feigenbaum was the only other researcher there studying chaos, but he left soon after and ended up at Rockefeller University. James Crutchfield, coming into his own as an innovator in his own right, moved on to a position at Berkeley, where he taught for a time and has continued to maintain an association. Out of the former "chaos cabal," only Robert Shaw ended up making his career in Santa Cruz, eventually working—many years later—at the Haptek Corporation.

Meanwhile bold new thinkers in the area of complex systems were coming into their own. At the University of Pennsylvania, a young biophysicist named Stuart Kauffman was engaged in experiments attempting to discern the origins of complex life. He developed a theoretical model by which animate qualities emerge from interactions of inanimate elements. Chris Langton, a graduate student at the University of Michigan, was exploring similar issues, attempting to create, as he called it, "artificial life." There, he enjoyed the supportive atmosphere provided by computational pioneers such as John Holland. Holland, the first person in the country to obtain a computer science Ph.D., had established at Michigan a laboratory for modeling the competitive and regenerative aspects of living behavior through machine code.

With the revolution in complexity emerging in places as far away as Berkeley, Los Alamos, Philadelphia, and Ann Arbor (Michigan)—let alone in Germany, Japan, the United Kingdom, and other places where similar projects were going on, collaboration proved difficult at first. Field leaders yearned for a neutral place where they could get together—perhaps during research leaves or sabbaticals—and toss around ideas.

In 1984, attempting to fill this gap, Murray Gell-Mann, along with nuclear chemist George Cowan, former head of research at Los Alamos, spearheaded an effort to found a research center in Santa Fe, New Mexico, dedicated to the investigation of complexity in its various aspects through interdisciplinary exchange. Gell-Mann saw the Santa Fe Institute as a place for unbridled discussion of novel ideas in a relatively isolated setting. The curly-haired Nobel laureate hoped that it would stimulate a scientific renaissance comparable in spirit to the heady days in which modern particle models were formulated.

As Farmer has pointed out, Gell-Mann was the only prominent particle physicist of his generation to have bridged the gap between the cutting-edge issues of the mid-twentieth century—namely particle dynamics and quantum field theory—and those of the late twentieth century—chaos, complexity, self-organization, and emergence. Gell-Mann's indisputable credentials as developer of the quark model—one of the greatest predictive leaps in modern physics—lent the institute considerable respect from the very beginning. His stature helped draw other exemplary researchers out to Santa Fe.

Two other Nobel laureates—Princeton physicist Philip Anderson and Stanford economist Kenneth Arrow—were among the first to become involved, joining Gell-Mann, physicist David Pines, and others. In short order, Holland, Kauffman, Farmer, Crutchfield, Packard, and Langton all became Santa Fe participants—either as visiting scholars or permanent staff. In the years that followed, dozens of other professors, postdocs, and students similarly found their way to Santa Fe, staying for a time, enjoying the commerce of ideas, and then moving on to other venues.

Within the sun-baked adobe walls of the institute, framed by beautiful southwestern desert vistas, a new predictive paradigm has come forth. Known as "emergence," it states that the effective behavior of a system is often much more than just the actions of its components. For example, living beings certainly act in a far more complex manner (acting independently, and, in the case of humans, consciously) than one would expect from just a collection of chemicals. Economic systems often display general trends that analysis of their working parts would not predict. And large-scale emergent physical behavior—crystallization of liquid water into complex snowflake patterns, for instance—is ubiquitous in nature. Considering the collective, interdisciplinary accomplishment that characterizes the Santa Fe Institute, perhaps nowhere else could such an innovative model for aggravate behavior, spanning practically all fields of study, be addressed in all of its wondrous aspects. Let's look at some of these achievements, and how they pertain to the modern science of prediction.

6

THE BODY ELECTRIC

Complexity and Living Systems

> *The advance from the simple to the complex,*
> *through a process of successive differentiations*
> *is seen alike in the earliest changes of the*
> *universe to which we can reason our way*
> *back; and in the earliest changes we can*
> *inductively establish; it is seen in the*
> *geological and climatic evolution of the Earth,*
> *and of every single organism on its surface; it*
> *is seen in the evolution of humanity . . . and of*
> *all those endless concrete and abstract*
> *products of human activity which constitute*
> *the environment of our daily life.*
>
> —HERBERT SPENCER
> *(Progress: Its Law and Cause)*

Life, but Not as We Know It

An odd-shaped creature, fashioned of unit digits in a sea of zeroes, approaches its helpless prey. Closer and closer it comes, finally arriving at the square-shaped bait. The entity absorbs the new material into its own form, growing bigger in the process. But alas, several competitors then arrive on the scene, all at the same time. The overcrowding proves too much, and all of the beings—including the original—instantly die, vanquished into pure nothingness.

151

Machine models of biological systems constitute a significant segment of the research taking place at the Santa Fe Institute, the Artificial Intelligence Lab at MIT, the Cognitive Sciences Department at Sussex University in England, and other major centers for computational research around the world. Some of these simulations involve actual robots, hard-wired and programmed to mimic living creatures. Others are "virtual realities," existing only on the computer screen. Their digestion, procreation, locomotion, and other functions proceed by means of transient interplays among computer bits.

There are countless reasons why scientists find lifelike models intriguing and important—including applications to speech systems, pattern recognition schemes, mobile robots, and so on. For those attempting to model the future and unravel its mysteries, they provide a special kind of gift—a way of learning from nature how it enriches itself through variation and organizes itself through the survival of the fittest. Researchers hope to simulate these attributes through computational algorithms, employing these features (among other applications) to build more efficient programs for optimization and forecasting—that is, modeling the best and most likely outcomes of an evolving system.

Supplementing and, in many cases, moving beyond standard statistical techniques such as linear regression (finding the line that best fits a set of data points) and autoregression (taking a weighted average of past data to anticipate future trends), biologically based prediction methods interpret and extend complex sets of information such as electroencephalogram data, stock market figures, and seismographic readings by exploiting the remarkable flexibility of natural systems. In recent years, science has developed a wide variety of such approaches. Genetic algorithms, which are basically optimization techniques modeled on chromosomal behavior during reproduction, employ Darwinian struggles between competing possibilities to find the fittest solution to a problem. Often, the optimal solution found constitutes a good predictor of future behavior. Neural networks, another computational method often used in forecasting, interpret and then extrapolate sets of information by means of analogies with the brain's ever-changing systems of interconnected neurons. These learning algorithms model the brain's remarkable ability to process information in parallel (many processes simultaneously), and to apply what it learns in gauging the likelihood of future possibilities. Yet other computational techniques, including evolution strategies, evolutionary programming, simulated annealing, and state-space analyses make use of additional natural features in mapping what is known and anticipating what might happen next.

Biological simulation dates back to the earliest days of computer science, when John von Neumann proposed modeling living organisms as automata: mechanistic processes following well-defined rules. In 1948 he gave a famous lecture in which he pictured robots programmed to replicate themselves by assembling their "progeny" from spare parts. He imagined generations of self-reproducing automata equipping their offspring with more and more advanced features. In a Darwinian process based in metal, not flesh, the custom-made robots would compete with each other for scarce resources—a struggle in which only the fittest would endure.

The Polish mathematician Stanislaw Ulam, after hearing about von Neumann's idea, convinced him that there was an easier way to perform such simulations. Ulam proposed to him the notion of cellular automata: organisms modeled as interacting site values on a grid. Taking up Ulam's suggestion, von Neumann designed a Turing machine–like system featuring an array of 0s and 1s updating itself periodically according to simple, local rules.

One might picture a cellular automaton as a peculiar game of checkers. Imagine a checkerboard in which each square is either empty or contains a single black piece; presume all the red pieces are lost. (In an actual computer model each "cell" would contain either a 0 or a 1.) Depending upon the nature of the simulation, the pieces are either randomly scattered throughout the board, or arranged according to some specific initial configuration.

Now the game begins. A single "step" or "generation" of the automaton involves a number of actions performed in synchrony. A program evaluates the contents of each square along with those of its nearest neighbors (the adjacent cells), and based on a preset table of rules, decides whether or not to alter that site—to retain, remove, or add a piece. For example, it might be instructed to add a black piece to each empty square surrounded by an odd number of checkers, but leave alone an empty square bordered by an even number. This process occurs again and again, as long as the algorithm runs.

One might observe a number of instinct types of cellular automaton behavior. If, at some point, all of the cells are empty and remain empty, one says that the population has "died out." If, on the other hand, a pattern of black squares remains unaltered by the rules from generation to generation, one refers to it as "stable." Other possibilities include periodic configurations, cycling though a fixed set of patterns and then back again—like a television being "flipped" through all of its channels until the original appears—as well as chaotic behavior. One typically cannot predict what will happen until one runs the actual automaton and examines its results.

Cellular automata formed an obscure part of computer science until in 1968 a clever young student at Cambridge, John Conway, designed an intriguing set of rules, which he dubbed the "Game of Life." His aim was to model how living systems function—representing beings and their food supplies (other beings) as patterns of 1s surrounded by seas of 0s. He purposely selected dynamical principles that avoid tendencies either to die out or to grow unchecked, balancing addition and deletion of elements. In this manner he increased the odds of interesting, long-lasting behavior.

Conway stated his rules quite simply and unequivocally:

Life occurs on a virtual checkerboard. The squares are called cells. They are in one of two states: alive or dead. Each cell has eight possible neighbors, the cells which touch its sides or its corners.

If a cell on the checkerboard is alive, it will survive in the next time step (or generation) if there are either two or three neighbors also alive. It will die of overcrowding if there are more than three live neighbors, and it will die of exposure if there are fewer than two.

If a cell on the checkerboard is dead, it will remain dead in the next generation unless exactly three of its eight neighbors are alive. In that case, the cell will be "born" in the next generation.[1]

In a 1970 column in *Scientific American,* Martin Gardner detailed Conway's idea and described some of the results obtained from running it on computers. He painted a curious world of strange animate creatures—dubbed "blinkers," "traffic lights, "toads," and so on, because of their resemblance to mundane objects. Most intriguing were "gliders": patterns that seemed to walk across the screen as the automaton carried out its steps. Reportedly, thousands of computer-adept readers of Gardner's column took breaks from their own projects, and spent many enjoyable hours testing the game out for themselves.

Mentioned in Gardner's column was a challenge posed by Conway. Conway offered $50 to anyone who could design a set of Life rules that would generate unlimited quantities of patterns. R. William Gosper of MIT's Artificial Intelligence Lab responded with an idea for a "glider gun": a configuration that would shoot out one glider after another. Conway gleefully sent Gosper the money, recognizing the profound implications of his discovery. The regularity of glider guns, Conway realized, could be used to assemble cellular-automaton-based computers—universal computing machines of the kind proposed by Turing. In other words, given the proper initial configuration, Life could be employed in solving any step-by-step calculation.

In the late 1970s and early 1980s, researchers began to realize that cellular automata represented much more than just games. Italian physicist Tomoso Toffoli initiated a critical examination of how these systems might be used to simulate fundamental physical behavior—such as crystal growth and phase transitions (transformations between gaseous, liquid, and solid states). Norman Packard cleverly devised rules that would produce a variety of patterns bearing a marked resemblance to snowflakes. MIT professor Ed Fredkin pondered the radical notion that the universe itself was a massive cellular automaton, generating everything that we see around us through simple rules—an idea that he tried to sell to his good friend Richard Feynman. Reportedly, Feynman was fascinated by Fredkin's concept, but—given the locality of cellular automata and the nonlocality of quantum dynamics—didn't believe it could work.

For a time, though many worked in the field, one man seemed to dominate: Stephen Wolfram. In a span of several years, Wolfram immersed himself in comprehensive studies of different types of cellular automata, producing a classification scheme for them as universally accepted as the MPAA rating system is for American movies. He divided automata into four broad categories, based on their observed behavior:

Class I: All activity eventually dies out; nothing remains
Class II: Automata evolve into stable or simply periodic patterns

Class III: Activity continues indefinitely, jumping through various states in a seemingly random manner

Class IV: Equivalent in behavior to a universal Turing machine (the Game of Life falls into this category)

The last grouping, IV, constitutes the smallest, but most interesting, class of automata. It is the only category that embodies self-sustaining locomotion, in the form of gliders, so-called "puffer trains," and other ambulatory features. It is neither simply periodic, nor purely chaotic, but, rather, situated on the cusp between these extremes. Consequently, researchers examining emergent, self-organized behavior in automata have made this fourth class the special focus of their studies.

The Glass Bead Game

In Hermann Hesse's masterful Nobel Prize-winning novel, *The Glass Bead Game,* a future society of scholars devote their time to developing connections between artistic, musical, and philosophical ideas—upholding the ancient Pythagorean ideal in times to come. The most esteemed citizens are those with the ability to discern patterns woven from disparate sources.

John Holland has always enjoyed solving a good puzzle—finding ways of making sense of competing notions and complex situations. For him, the computer is not just a device for calculation, but a tool for decoding the innumerable enigmas of nature. Since the 1960s, he has engaged in a sometimes lonely quest to incorporate the collective and evolutionary features of living systems into dynamic computational models. It is hardly surprising that *The Glass Bead Game* has remained his favorite book; indeed, at times when his work has been not as well understood, it has served as a source of inspiration and comfort.

Born in Indiana in 1929, Holland grew up in a small town in western Ohio. A strong student, excellent in math and physics, he won a scholarship to MIT during the critical period in which computers were first being developed. Notable founders of the field of computer science, such as mathematician Norbert Wiener, roamed MIT's "infinite corridor" (a hallway that seems to stretch from Massachusetts Avenue to the great beyond), seeking converts to the notion that

FIGURE 6.1 John Holland (1929–), founder of the field of genetic algorithms and pioneer in the study of emergence (courtesy of the University of Michigan).

machines and living organisms have much in common. Wiener's idea, called "cybernetics," stimulated a generation of interest in robotics, and inspired Holland and others to look for deep connections between computational devices and biological systems.

After obtaining his undergraduate degree, Holland wisely decided to continue his education at the University of Michigan, where he had the opportunity to study with philosopher Arthur Burks. For a philosopher, Burks leaned strongly toward the applied side. He worked with von Neumann on designing some of the earliest computers—initially applying his talents to the very first computer, the ENIAC at the University of Pennsylvania, then working on its more sophisticated successor, the EDVAC. After von Neumann's premature death in 1954, Burks sought to carry on his legacy by instilling in its students an appreciation of the profound importance of viewing computers not just as practical machines but also as means for understanding the mind.

Under the mentorship of Burks, Holland became even more fascinated by the prospects of modeling biological systems and drawing analogies with computation. As Holland describes that period: "It all started when I was browsing in the Math Library . . . and encountered Fisher's book, *The Genetic Theory of Natural Selection.* That one could do serious mathematics about the mechanics of natural selection (a topic that had always interested me) was a revelation."[2]

Holland received his Ph.D. in 1959. Encouraged by Burks, he remained at Michigan as a postdoctoral fellow, then joined the faculty, receiving tenure in 1964. By then, he had given considerable thought to the concept of developing computational models for two natural processes that had eluded full explanation: human learning and genetic adaptation. Holland insightfully realized that these distinct processes might be viewed as manifestations of a continuum—dual faces of the same head of Janus.

According to Holland, the ability of the brain to acquire and utilize knowledge and the propensity of genetic material to organize itself indicate that they are each examples of what he calls "complex adaptive systems." Additional instances of such systems include social organizations and the economy. In fact, as Holland points out, once one recognizes their features, one finds them everywhere.

Holland defines a "complex adaptive system" as having four critical aspects.[3] First of all, it must consist of a multitude of distinct components acting at the same time. In the case of the brain, these parts are the neurons; for living organisms, the genes; for organizations, the members and member institutions.

Second, any complex adaptive system must have multiple layers of organization, with agents on each level serving as building blocks for the next. For example, in industrial organizations, each worker might belong to a unit, which, in turn, might function as part of a division, which itself might be part of a larger company, and so forth. In ecological systems, similar roles are served by individual organisms, groupings, and habitats. These organizations continuously modify themselves, as necessity mandates, forming new links and clusters.

Third, a functioning complex adaptive system never reaches a state of equilibrium with its environment. Ice cubes placed in a glass of hot water will eventually melt—the entire contents achieve a compromise temperature. A living entity cannot afford to make such a deal with the disordered world around it. It maintains its own internal order by pumping entropy out into the surroundings (that is, taking in nourishment and giving off waste). In short, if it ever reaches equilibrium, it is dead.

Finally, and most important, complex adaptive systems must have the capacity to anticipate the future. Not that organisms, neurons, and so on are fortune tellers. Rather, they need to continuously interact with the outside world, use the information they gather to develop predictive models, and adjust their own behavior accordingly. Otherwise they are at a competitive disadvantage.

This process need not be conscious. Humans make decisions based on their own awareness, but more primitive organisms lack such capacities. Rather, they store their knowledge in their genes. As they evolve, their genetic material helps them respond to their environments. Systems with superior genes gain a competitive advantage, passing along this advantage to their progeny. Conversely, ill-adapted genetic combinations die out over time. In this manner, useful knowledge of an organism's surroundings becomes encoded in its hereditary makeup.

Peak Performance

One of Holland's most important contributions to the field of computer science is the notion of genetic algorithms for optimization. By taking advantage of the adaptive properties of genetic material, his technique is a powerful way of searching for the optimal solution to a problem. Since the optimal response is often the one carried out, optimization techniques often provide excellent predictions of future behavior.

For example, a woman walking across an empty field would most likely take a fairly direct path, reasonably close to a straight line. This optimal route would thereby constitute the best forecast of her actions. She could do otherwise, of course, but, chances are, she'd select an expedient way of crossing.

Optimization, in general, is a branch of mathematics concerned with exploring ranges of possible outcomes—known as "fitness landscapes"—and finding out which eventualities are most suitable. An optimization routine attempts to hone in on the fittest possible solutions within a given landscape. Holland's method utilizes principles of natural selection to do so in an efficient manner.

The traveling salesman dilemma is one of the best known optimization problems. Given a set of cities to be visited by a salesman on his route, which is the best order for him to visit them? The "fitness landscape" of this problem consists of the set of all possible routes and their lengths. The points on the landscape with greatest fitness correspond to routes with minimal length. They are known as the "fitness peaks"—the Himalayas of the terrain. Other regions of the landscape, corresponding to less favorable routes, are called "fitness valleys." The question is: If a computational routine for solving the problem offers a suboptimal solution—located, say, in one of the valleys—how might it most efficiently proceed to the highest peak?

Unless they have multiple processors, standard computers must examine each option in serial fashion. Therefore, with larger and larger numbers of cities, resolution takes longer and longer, until the wait becomes intolerable—eventually even astronomical. Genetic algorithms considerably reduce the amount of processing time. In this method, each possible solution to a problem is represented by a unique string of 0s and 1s. The researcher defines a function that evaluates the fitness of each string. In the traveling salesman problem, for instance, the fitness function determines the shortness of the route.

Now here's where the biology comes in. Each time step, half of the strings, acting as "chromosomes," engage in the genetic process of "crossover" with the other half. Crossover in biology means that two chromosomes from mating organisms exchange genetic material. This procedure, also known as recombination, is a natural part of reproduction, allowing offspring to harbor characteristics of both parents. In Holland's model the same thing happens with the binary strings. They split at random points, and swap parts—the first segment of one merging with the second segment of the other, and vice versa. Some of these "offspring" then receive "point mutations": small random changes in their composition (0s turning into 1s, or the opposite). Finally, whichever children are fitter than their parents replace them in the genetic pool.

Here's an example. Suppose two mating strings, randomly selected from a set, are 000111 and 101010. In the crossover process they each split down the middle, becoming: 000–111 and 101–010. They exchange material, producing: 000010 and 101111. The first offspring receives a point mutation, becoming 100010; the second remaining 101111. Based on the values of a predetermined fitness function, the two are deemed fitter than their parents, and thereby become part of the next generation of the population. This process continues, again and again, until the fittest combination of binary digits emerges, representing the optimal solution to the problem under consideration.

Holland has chosen these rules for a number of reasons. Crossover, he argues, offers the possibility of favorable segments of genetic material being passed down from generation to generation, and unfavorable ones being discarded over time. This, he feels, leads to faster movement toward fitness peaks. Mutation, he contends, helps avoid the conundrum of being stuck near false optima—peaks of high, but not the highest, fitness. It randomly jostles strings into other regions of the landscape, allowing them to reevaluate their course, and move (one hopes) toward the absolute highest peak.

Holland's is not the only optimization routine based on biology. In recent decades, other scientists around the world have proposed a wide range of alternative methods. For example, the evolution strategies technique, developed in the 1970s by German scientists Ingo Rechenberg and Hans-Paul Schwefel, uses single values, instead of strings, to represent conceivable solutions. Recombination takes place by means of averaging these numbers, and mutations occur by means of random changes in value (selected to have a special distribution). Evolutionary programming, a similar method proposed by Lawrence Fogel in the 1960s, mimics evolution on the scale of species, rather than individuals. Consequently, it refers to organisms, rather than chromosomes, and, in contrast to the other two techniques, does not include recombination. Researchers refer to the entire class of methods as evolutionary algorithms.

Other optimization techniques are based on physical, rather than biological, principles. For example, the method of simulated annealing borrows thermodynamic techniques to find optimal solutions to problems. In thermodynamics, atoms jostle around a lot at high temperatures, but move about very little when temperatures are low. Similarly, in finding the fittest answer to a particular query, one tends to probe a wide range of solutions during the start of the search, but then focus in on a small subset as the search nears its conclusion. In simulated annealing, the abstract process of closing in on the fittest solution is modeled by the physical process of lowering a system's temperature and freezing it in place.

A typical simulated annealing routine involves boosting a fitness function (of several variables) residing in a suboptimal state up to its optimal value. For example, in the traveling salesman problem, such an algorithm would start with a route that was less than ideal, and end up with the best route possible. To perform this task, the algorithm searches through the fitness landscape, replacing the original parameters with slightly different values. In the case of the traveling salesman, this would involve making small changes to the route. If the fitness function increased on such a substitution, then the new solution would be better than the old (a shorter route, for instance). If it decreased, then the new solution would be worse (a longer route).

In many optimization algorithms, better solutions would automatically be retained, and inferior solutions immediately discarded. This could well lead to a situation, however, where a routine becomes stuck on a suboptimal peak, unaware that better solutions exist in a wholly different sector of the physics landscape. Imagine, for example, a traveling salesman planning a route from Los Angeles to Orlando by way of Chicago, then Detroit, then Denver. Rearranging options in his mind, he might at first think that such a route is optimal, because it is better than going to Detroit first, then Chicago. Only by considering a wider range of possibilities would he eventually conclude that going to Denver first (from LA) would be best.

To accommodate the need for the broadest possible outlook in the beginning of a search and a narrower focus as the search winds down, simulating annealing sets the "temperature" of change high at the start, then gradually lowers it. During the "hot" era, the routine liberally scans through all possibilities, rarely rejecting any. Hence, like a desert nomad, it haphazardly explores a large region of the fitness landscape. Then, as the system "cools," the algorithm grows pickier. More and more, it only permits changes that bolster overall fitness. Because one hopes that by then the program has come close to finding the final answer, radical changes no longer help. Finally, like a crystalline ice formation, the system "freezes" in place—presumably in its optimal configuration (a traveling salesman's shortest route, for instance).

Optimization techniques represent great aids for predictive science. By determining the best approach to a particular issue, one might anticipate most likely outcomes—assuming that others are engaged in optimal strategies as well. A poker player, knowing the best ways to play a hand, would likely outperform less educated opponents, not just because he could make better moves himself, but also because he could better predict their responses. In other words, his experience would enable him to form models in his own mind of what others might do—useful in deciding what to do next.

A Child's View of Learning

A five-year-old is baking a cake. She climbs up into the pantry, finds some flour, sugar, cocoa, chocolate chips, and other assorted ingredients, mixes them together in a bowl, and places the concoction in the microwave oven. All the while, she pictures how wonderful the cake will be. She thinks about how fun it will be to taste all the chocolate. She imagines her family eating the cake and telling her how delicious it is. Finally she envisions her mother smiling at her and telling her how proud she is.

Suddenly, the smoke alarm goes off, and her vision bursts. She opens the oven, and finds a burnt, bubbling mess. Her mother runs in and screams. Instead of enjoying a homemade dessert with her family, the child, now miserable, spends the rest of the afternoon up in her room.

The next day, her parents help correct her misperceptions. They patiently show how her how cakes are made. She updates her mental image to include all the steps and ingredients she left out. They describe to her the dangers of improperly using an oven, and she solemnly thinks about what worse things could have happened. Consequently, she becomes much more careful, and—until she is much older—never tries to bake unsupervised again.

To predict what will happen next in the world often requires the construction of detailed mental models. These are internal representations of external reality, assembled on the basis of our sensory information and experiences. By manipulating these models in our minds, we are able to imagine various possible futures, and subsequently act on the basis of these conjectures. However, because these models are merely representative, in many cases they turn out to be inaccurate. As noted Princeton psychologist Philip Johnson-Laird has remarked:

> At the first level, human beings understand the world by constructing working models of it in their minds. Since these models are incomplete, they are simpler than the entities they represent. In consequence, models contain elements that are merely imitations of reality—there is no working model of how their counterparts in the world operate, but only procedures that mimic their behaviour.[4]

Because the accuracy of our predictions typically depends on the validity of our mental models, learning about the world forms a critical aspect of forecasting. We often learn best when our mistaken conjectures clash with what new experiences tell us. In that case, we update our internal representations accordingly. Through a continuous process of trial and error, our mental models increasingly represent a greater and greater range of aspects of the world. Yet, by their very nature, they can never encompass everything; they only reflect, in a limited way, the fraction of the world's features that we manage to filter through our senses and process in our brains during our lifetimes. Each of us hopes, however, that the scope of his own mental model offers enough predictive powers for him to lead a successful life.

When the computer was first developed, many of its designers hoped that it would someday be able to reproduce salient aspects of human intelligence. The Artificial Intelligence (AI) movement, founded by computer scientist Marvin Minsky and others at MIT, has aspired to show that computer hardware and software will eventually be powerful enough to duplicate every demonstrable ability of "wetware" (the human brain)—from pattern recognition and language processing to consciousness itself. Other thinkers, such as Roger Penrose, have contended that certain aspects of the mind are unique to humans (or at least to living cognizant beings) and cannot be precisely reproduced, only vaguely imitated. Regardless of their stance, computer scientists agree that computers can be programmed to model at least *certain* features of intelligence.

Consider what a machine would require to make useful predictions. Based on human experience, it would certainly need an internal representation of the world—at least of the data relevant to the forecast being rendered. This could be programmed in directly, through a set of values and equations. For example, a baseball prediction machine could be supplied with the initial position and velocity of a baseball, as well as the formula for projectile motion.

If, on the other hand, we wanted the machine to emulate human thought processes in a truer sense, we would withhold all formulas, and force it to learn them on its own. We would train the machine by offering it data, seeking its response, and then instructing it on whether or not its answers are correct. Thus, in theory, we could teach it any algorithm. Once its internal model became well developed, we would feed it novel information, and then ask it for predictions. In short, we would treat it like a child acquiring new concepts through trial and error.

To learn like a child, a machine requires a childlike brain, one that is as flexible as possible. Whenever children learn new facts or discover new associations between ideas, the neurons in their brains alter the strengths of their various connections. In fact, researchers believe that the plasticity of the brain's neural pathways permits novel ideas to be generated—in adults as well as children.

Connectionist models of learning seek understanding of these higher-order processes by examining how organization emerges from the local actions of linked elements. As complex adaptive networks, they are closely related both to optimization strategies and to cellular automata. Because they attempt to model the behavior of the brain's neurons, they are often referred to as "neural networks." This expression is somewhat misleading, however, because model neurons, as elements of connectionist models, act in a far more limited way than actual neurons. Nevertheless, these approaches represent highly successful computational strategies for acquiring detailed knowledge and making enhanced predictions.

A typical connectionist system consists of a network of linked processing units (representing neurons), running in parallel. The network is furnished with information about a procedure, then asked to react. Given this data, each processing unit responds with a different output, competing with each other to produce a definitive response. Connections are assigned different weighting, mandating that the conclusion reached relies more on certain processing units than others. The mechanism then evaluates the validity of its results, adjusting the weighting of its connections to improve on its performance. In other words, the system corrects itself more and more—in a continuous feedback loop—until it "learns" how to mimic the procedure correctly.

One of the simplest feedback algorithms derives from a suggestion Donald Hebb made in 1949 in his book *Organization of Behavior*. Hebb's rule states that the more a connection is used—as determined by the activation of neurons on both ends—the more its weighting should increase. This ensures that processors judged more useful become "listened to" more than those that prove worthless.

In the 1950s, 1960s, and 1970s, computer scientists in Europe, Japan, and the United States explored various connectionist models. Researchers such as Christoph von der Malsburg, Teuvo Kohoven, Shun-ichi Amari, Bernard Widrow, Marcian Hoff, Stephen Grossberg, and others developed computational representations of perception, pattern recognition, attention, memory, motivation, motor control, and other functions of the brain. Unfortunately, at that time few outside the field were aware of the work being done. Predictive experts, on the whole, had yet to find use for connectionism, viewing it as an abstraction.

All this changed in the late 1980s, when young devotees, who had already embraced the mantras "chaos" and "complex," took up the chant of "connectionism" as well. As buried achievements of the 1960s by gurus Lorenz, Holland, and others became unearthed, machine learning research was next to follow. Neural networks suddenly became recognized as a hot "new" topic of study.

A remarkable theoretical breakthrough in connectionism, a powerful new method called backpropagation stoked additional fuel into this dynamo of resurgent interest. Invented in 1974 by Harvard Ph.D. student Paul Werbos in his thesis "Beyond Regression," and independently rediscovered in the mid-1980s by researchers David Parker and Yann LeCun, backpropagation helped make neural networks much more flexible and efficient learners. Unlike previous algorithms, it permitted machines to think backward as well as forward, allowing them to reevaluate all their premises.

Reevaluation is an indispensable tool of learning and prediction. Often we are forced to completely rethink our assumptions. For example, imagine having a conversation with a society lady in which she continuously refers to someone named Hubert. "I took Hubert out to Chez Albert last night," she says. "It was quite an occasion. He very much enjoyed the truffles." All the while, you form a mental image of Hubert as a wealthy gentleman, probably her husband.

Perhaps you make a prediction based on this assumption. "I imagine Hubert will be driving you to the Charity Ball next week."

She laughs and replies: "I'm leaving him in the kennel for that one."

After momentarily imagining her upper-crust husband in a kennel, suddenly you realize that Hubert is a dog. Embarrassed, you review everything you said about Hubert, and completely reconstruct your mental model of him. You venture another conjecture based on your revised hypothesis:

"Well I guess he'll be eating Puppy Chow instead of steak."

Her affirmative nod indicates that your image of Hubert is now correct. By rethinking your suppositions, you have enabled yourself to make a successful prediction.

Computers, like humans, might get off track. When data is incomplete, they might well leap to wrong conclusions. Early neural networks did not fully take this possibility into account. After venturing a hypothesis and having it validated by a few examples, they treated all new information as refinements. If something completely different came along, they could not look back at prior data through the lens of revised assumptions. Rather, they were forced either to keep the existing model, flaws and all, or to completely start from scratch with the new data.

For instance, if a sequence started with "2, 4, 6, 8," an old-style neural network might learn it to be an ordered set of even numbers and anticipate that the next would be "10." If the series continued with "9, 7, 5, 3, 1," however, it would either switch its assumption from evenness to oddness or forever make bad guesses. It wouldn't look backward at its earlier data and conclude, as an alternative, that the set consisted of all the numbers between "1" and "9" in some more subtle arrangement.

A neural network with backpropagation, on the other hand, remembers everything and continuously updates its hypotheses on the basis of all of its available information. It consists of three distinct parts: input nodes, where novel information arrives, output nodes, where calculated estimates emerge, and hidden nodes, the meat of the sandwich, where comparisons take place between computed values and actual data. Each of these nodes is linked to the others by means of weighted connections.

During the training phase, the program initially assigns all the weights arbitrary values. Then the hidden nodes get to work, processing the input values. After computing the weighted sum of the information it receives through its connections, each node passes on the result to its neighbors. Ultimately, some of the hidden nodes transmit values to the output nodes.

Next comes a comparison between the calculated output values and actual information. Based on the difference between the two, the output nodes inform the hidden nodes about how best to minimize the error. This involves adjusting each weight throughout the system in such a way that the overall error goes down.

After this undertow of backpropagation proceeds, the waves of calculation move out again, generating a new set of output values. Once more, these are compared with the actual amounts. To reduce the net difference, the weights of all the nodes are again changed accordingly.

Again and again, these forward and backward processes occur, until the model's output and the genuine values match. Once that happens, the algorithm stops and the weights of all the nodes are recorded. The machine has finally learned the task and can model new sets of data on its own.

Neural networks can be applied to a vast range of learning situations, from game playing to pattern recognition, from analyzing seismographic records and predicting the likelihood of earthquakes to examining the behavior of the stock market and forecasting future trends. Naturally, the simpler the system, the faster and easier it is for a connectionist device to realize how to master it.

In a science museum, I once played several rounds of tic-tac-toe with a basic neural network learning system. At first, the machine lost a number of games, while it was testing random responses. But then, as time went on, it adjusted its internal weightings to allow for more effective moves. Soon, it was winning, or at least forcing a draw, every game.

For more complex analyses, machine learning algorithms often require more time and guidance. As the Turing Halting Problem has shown, some tasks would stump any conceivable machine. Others would take so long that the wait would hardly be worth it.

Under normal circumstances, however, neural networks perform remarkably well. By modeling the decentralized functioning of the brain, these parallel-processing techniques manage higher-powered acquisition of knowledge than ordinary serial procedures can typically obtain. Once again, biology proves a great source of computational ideas.

The Edge of Chaos

Why do some computational systems possess higher-order predictive properties, and others simply churn out nonsense? Why do some combinations of chemicals yield self-perpetuating living beings, and others result in lifeless broths? How did organized intelligence evolve?

Chris Langton has made it his life's goal to map out connections between these possibly related questions. Fascinated as a youth by the Game of Life, he believes that their answers lie in an understanding of cellular automata and related systems. Life, intelligence, and other emergent features of nature, he feels, are manifestations of special kinds of dynamics—neither rhythmically simple nor haphazardly complicated, but rather somewhere in between. He calls this critical region "the edge of chaos."

Langton, a native of Massachusetts, was one of the many graduate students of Arthur Burks and John Holland at the University of Michigan. He shared with his mentors an interest in connecting computers with biology, so much so that he was often urged to veer more toward the computational side. Despite pragmatic concerns, such as getting his degree, his enthusiasm for examining living creatures through the virtual microscope of computer programs was unbounded. To describe his vision, he coined the term "artificial life," detailing this concept to Doyne Farmer when they happened to meet at a 1984 conference at MIT.

Farmer was intrigued by Langton's notion, and invited him out to the Center for Nonlinear Studies at Los Alamos for several seminars and workshops. At one of these, in May 1985—attended by Farmer, Packard, Wolfram, and others—Langton presented his ideas about cellular automata and the edge of chaos. Reportedly the attendees were extremely impressed. Farmer, who around that time became the leader of T-13, the complex systems group at Los Alamos, offered Langton a postdoctoral fellowship.

The essence of Langton's research concerned a way of treating Wolfram's cellular automaton classification system as a multiphase model—comparable to categorizing water in solid, liquid, and gaseous forms. Just as liquid water organizes itself into crystalline structures as its temperature is lowered, and boils away as its temperature is raised, Langton proposed a special parameter, called λ, that alters the qualitative properties of automata. For an automaton with a given set of rules, he defined λ as the fraction of site values the rules transform into ones. Using the terminology from the Game of Life, one might say that λ is the percentage of cells that are alive, on average, after a given step. For Class I automata, where everything quickly dies out, λ is very small. For Class II, where sites evolve into stable or simply periodic states, λ is somewhat greater. In the case of Class III, or chaotic automata, where sites transform into 0s and 1s, and back again, with roughly equal odds, λ is close to 50 percent.

Like a chemist experimenting with raising and lowering the temperature of a reaction, Langton altered lambda up and down, examining the effects. He observed a marked phase transition, right at the cusp between Class II and Class III. Noting the "magic" value of lambda for this qualitative shift—roughly .273—to Langton's glee, he found that it neatly corresponded to Class IV automata, namely the Game of Life rules. In other words, the complex, self-perpetuating world of the Game of Life—with the ability to engage in Turing machine–like computation—lay smack on the border between the realm of pure order and the realm of pure chaos. Life literally seemed to reside on the edge of chaos.

An independent result by Packard seemed to corroborate Langton's findings. Packard ran a genetic algorithm simulation testing which cellular automaton rules produce the most efficient computational machines. The algorithm honed right in on Class IV automata as the best for solving problems. Because genetic algorithms model natural selection, Packard's results seemed to suggest a Darwinian process leading to particular modes of behavior, namely self-driven, "living" structures such as glider guns.

According to Langton and Packard, systems for learning and prediction, such as artificial neural networks and human brains, would operate best in the threshold region represented by Class IV. A set of neurons firing (sending signals) periodically—or conversely firing randomly—would scarcely be able to model the world's diversity and generate reasonable forecasts. Only one behaving in a complex but intentional manner—resting on the brink of chaos—would possess the richness of structure required to master the task.

Langton and Packard's work, as soon as it was published, almost immediately came under criticism. A young University of Michigan researcher from Holland's group, Melanie Mitchell, joined James Crutchfield in calling "edge of chaos" theories into question. Performing their work at the Santa Fe Institute, along with student Peter Hraber, they published several articles casting doubt on the idea of a special value of the parameter leading to particularly noteworthy behavior.

Mitchell, Crutchfield, and Hraber, however, emphasized in their papers that even though they found no evidence for special properties of the phase transition between order and chaos, they did not want to rule out future research on that subject. On the contrary, they argued that evolving cellular automata using genetic algorithms (as Packard did) was an endeavor deserving even more attention than it had received already. To that effect, they and several other researchers formed the Evolving Cellular Automata (EvCA) project, a major working group of the Santa Fe Institute that continues to this day.

Gene Nets and Artificial Pets

Stuart Kauffman, the exuberant, scraggly-haired, dungaree-wearing recipient of a 1987 MacArthur Foundation "genius" award, was the first to articulate the notion of antichaos—the process by which systems organize themselves from simpler to more complex forms. He focused particularly on the emergent properties of gene nets, also known as Kauffman nets or random Boolean networks. His model imagines a set of genes, each of which can be either active or inactive, interconnected by a set of random links. Simple automaton rules determined by the state of a gene and its neighbors (genes to which it is connected) determine whether or not it maintains its level of activity from generation to generation.

Kauffman envisioned these linked genes as a set of lightbulbs, randomly wired together, successively switching each other on and off. A darkened bulb in such a network, for instance, might flicker on for a moment, then off again, depending on how many bright bulbs are linked with it at any given time. The effect is akin to a Christmas tree, draped with lights programmed to turn on and off at set intervals. Only in this case, the rhythms of the glowing bulbs are set by the system itself.

Kauffman found, in running his model, an astonishing degree of ordered behavior emerging from what at first seemed a mere jumble. Perhaps, he speculated, primitive genetic systems organized themselves in this manner, evolving over time from meaningless chemical arrays to critical blueprints of life. Gradually, as life acquired greater and greater complexity, it learned how to process information about its environment—ultimately gaining conscious awareness and the ability to envision the future.

A related project called Tierra, designed by Tom Ray of the University of Delaware and the Santa Fe Institute, set out to model the evolution of life through a sophisticated computer simulation. Ray created a complex virtual landscape full of digital organisms programmed to replicate themselves, and ran the system forward in time to see how these creatures behaved. To encourage variety, he allowed for the possibility of mutations, which randomly changed organisms' instructions. The artificial creatures were given free rein over the computer's processors and memory—akin to the unwanted behavior of viruses.

Tierra's first run lasted but a few hours of processing time. Yet within that limited time span, Ray found considerable evidence of evolution toward more sophisticated digital beings. Some of the creatures mutated into parasites—entities dependent on others for survival. Other organisms learned to replicate more quickly by working together—discovering the advantage of a simple kind of sexual reproduction. All in all, Ray was impressed by the amount of diversity his simulation produced.

The construct of artificial habitats offers promise not just for understanding life's origins, but also for predicting its future. Ray and other ecologists are particularly interested in how diversity might be maintained in the face of environmental threats. As global warming, pollution, deforestation, and other manmade hazards impact countless species, ecological forecasting has become increasingly vital. Saving Earth's diverse forms of life requires unprecedented forms of planning—regulating waste disposal, limiting development in certain regions, and so forth—for which computer models can provide invaluable support.

Fractal Prognoses

Not all of the work in complexity concerns itself with simulations. Today, many researchers apply models gleaned from complex systems theory to direct studies of human patients. By using these theoretical methods to examine imaged portraits of bodily functions—obtained through EKG (electrocardiography), EEG (electroencephalography), MRI (magnetic resonance imaging), sonography, and various types of tomography—they hope to develop improved diagnostic and therapeutic techniques.

The importance of accurate, noninvasive diagnoses for certain deadly, rapid-onset medical conditions cannot be overestimated. Ventricular fibrillation, for example, a condition in which the chambers of the heart suddenly begin to pulsate in uncoordinated fashion, acting like a "bag of worms"—causes hundreds of thousands of sudden deaths each year in the United States alone.[5] In some cases physicians understand the cause: blocked arteries, hypertension, extreme stress, and other related factors. In other instances, though, they cannot identify a single reason for an obviously healthy person to suddenly die.

What if, during a routine physical, data from an electrocardiogram could be analyzed and used to identify a wide range of possible risks—including fibrillation, strokes, and other sudden-onset conditions—rating them according to their likelihood? Individuals with well-above-average chances of problems with their hearts or blood vessels could be given extra monitoring and the best available preventive treatment. Countless lives could be saved by timely intervention.

Ary Goldberger, a researcher at Harvard Medical School and Beth Israel Hospital, has spent much of his career advocating the application of complex systems techniques to diagnosing medical conditions, particularly heart problems. He and his colleagues have collected numerous gigabytes of EKG data, trying to identify the precursors of fibrillation and other kinds of arrhythmia (nonstandard behaviors of the heart). Their counterintuitive results, based on years of analysis, offer considerable promise.

Normally, we think of the healthy heart as a rhythmically beating organ, as reliable as a clock or a metronome. True, based on common measures such as taking a pulse, hearts do seem to beat very consistently. As Goldberger and his coworkers discovered, however, precise EKG plots yield a far more complicated picture. A normal heart, they found, tends to alter its beats subtly and aperiodically each time. Only when it is aged or diseased, they noticed, does it tend to beat in step.

To understand how hearts can seem regular on a coarse scale, but irregular on a fine scale, consider the analogy of a marching column of soldiers. Watching such a parade from a distance, one would be struck by the rhythmic cadence of their pounding feet and the pendulumlike consistency of their swinging arms. Yet, if one walked beside one of the soldiers as he trampled on rocks and soil of various textures, one would notice minor inconsistencies in each step. The same with a healthy heart; varying its rhythm slightly, it maintains, nevertheless, a steady overall pulse.

Goldberger speculates that a certain measure of chaos allows a well-functioning heart to respond in a more flexible manner under widely changing conditions. That way, it can ease into new rhythms, if necessary, without suffering too much strain. A diseased heart that beats with stricter periodicity, on the other hand, cannot accommodate stressful situations as well. Therefore, it is more likely to fail, or launch into severe arrhythmia, under such circumstances.

The EKG data collected by Goldberger and his colleagues constitute examples of what is called a time series. A time series is a record of the behavior of one or more response variables (experimental values) observed at fixed time intervals—for instance, a weather chart showing daily temperature changes. As a predictive device, it is indispensable; understanding the variations of past conditions offers considerable guidance for extrapolating into the future. Ideally, interpretation of a time series is aided by knowledge of causal or underlying factors. For example, knowledge of seasonal and climatic differences helps one appreciate the nuances of temperature charts. But often one does not know the reason for fluctuations, and must extrapolate solely on the basis of mathematical and statistical cues.

The mother of all statistical forecasting techniques is linear regression. In the simple case of a single response changing over time, this involves finding the straight line that best fits a set of data points. In this method, researchers begin by plotting a scattered cluster of information—temperature readings, for example. Each point's horizontal value represents the amount of time elapsed and its vertical value, the response at that particular time. Next, they use an algorithm to find the straight line that comes closest to each of the points. Technically, this involves using calculus to minimize the sum of the squares of each of the vertical distances between candidate lines and data points. They thereby obtain the parameters (slope and intercept) of the best fit straight line, which they can then use to try to predict future responses.

Imagine, by analogy, a newspaper delivery boy assigned to throw papers onto the lawns of a number of scattered houses (points). Unfortunately, the wheel of his bicycle is stuck in a way that he can only ride in a straight line. Therefore, to avoid getting a sore arm, he must choose his path carefully to minimize the distances between the path his bike takes and each of the houses on his route. Consequently, his path constitutes the straight line that best fits his objectives.

If time series data does not resemble a straight line, then statistical theory suggests use of nonlinear fits, such as matches to polynomials or other types of functions. The variation of average temperature throughout the seasons, for instance, best fits a sine wave, not a straight line. In modeling more than one response (for example, temperature and pressure versus time), then statistics offers the technique known as multiple regression: plotting the data in a space of as many dimensions as there are variables, and then finding the higher-dimensional model that best fits it. Each of these cases represent generalizations of basic linear regression.

But what if data appears random, like scattered thumbtacks on an unused bulletin board? What if its trend is either nonapparent or nonexistent? How might a scientist or statistician know what kind of line, curve, or other geometric object to select for an optimal fit? How might he distinguish information that is truly random from data with a hidden pattern—matchable in some way to a predictable model?

Statistics offers a number of options in such cases. One choice is to try out the widest possible range of models, hoping to stumble on the right match (namely, that fits the data as closely as one would like). But if the data turns out to be stochastic (truly or seemingly random, evolving over time in a probabilistic manner), such a search would likely prove futile.

Often a more sensible alternative is to eschew models altogether and let the data speak for itself. What sort of path into the future does the information itself suggest? In statistics, such methods are called non-parametric techniques: procedures for smoothing out, filling in, and then interpreting the shape of the raw data, rather than fitting it to a line or curve. Briefly, these procedures include averaging the data over fixed time intervals to make it less "bumpy," expressing it as a sum of its spectral components (sine and cosine functions, each of different height and spread), and finding its autocorrelations (special mathematical relationships between various data points indicating how similar or different they are). In the manner that a good makeup artist covers up wrinkles, smoothes out blemishes, and enhances attractive features, these routines help make the raw information "prettier"—hence easier to understand and extend.

Most time series forecasting today relies on tried and true algorithms—often combining a number of statistical "tricks." These are often variations of autoregression (AR), an approach introduced in 1927 by researcher G. U. Yule to predict annual numbers of sunspots. Without a working model of the Sun's behavior, Yule realized that he needed sophisticated mathematical means of squeezing the maximum amount of predictive information from past measurements. He developed a way of using the weighted sum of a set of observations to anticipate the next value. Applying this technique to sunspots, he generated reasonably reliable forecasts.

Yule's method, along with refinements suggested by Gilbert Walker, reigned supreme for almost half a century. Though popular, the Yule-Walker approach had clear limitations. Best suited for what statisticians call stationary (having no clear trend), stochastic processes, it could ill handle situations in which response variables displayed unidirectional tendencies—rising on average over time, for example.

In the 1960s and 1970s, George Box and Gwilym Jenkins combined autoregression with a number of other statistical techniques and developed a much more flexible procedure. In the manner of conscientious physicians, the good doctors offered careful diagnoses and a variety of treatments for a wide range of possible time series conditions. They distinguished stationary series, nonstationary series, and series with seasonal variations (displaying long cycles, like the seasons of the year), and provided distinct prescriptions for each. Respectively, these are called the ARMA (autoregressive moving average), ARIMA (autoregressive integrated moving average), and SARIMA (seasonal autoregressive integrated moving average) methods. Without getting into the mathematical details of these approaches, each uniquely combines "moving averages," a smoothing technique, with autoregression and other analyses. By designating various categories of time series, and outlining means of dealing with each one, Box and Jenkins helped render the field more systematic. Applicable to a variety of forecasting problems ranging from rainfall prediction to financial analysis, the Box-Jenkins procedure has since become the standard statistical approach.

Conventional statistics, however, falls short in modeling apparently random systems driven by underlying deterministic principles. As chaos theory as shown—and as Lorenz's butterfly effect and Shaw's slowly leaking faucet illustrate—many interesting systems behave in seemingly erratic fashions owing to their own internal dynamics. Standard statistical methods often cannot detect the patterns hidden within apparently haphazard data. Therefore, to model systems that are deterministic but chaotic, one must turn to more recent, specially developed approaches.

With the advent of enhanced methods stemming from complex systems theory, time series forecasting has taken a quantum leap forward. Supplementing contemporary statistical approaches, two new strategies in particular have launched the field into revolutionary new ground.

The first is the application of neural networks to time series. As we've discussed, by teaching a connectionist learning system to model a set of data, programmers might then ask it to extend the model into the future. Economists, as well as medical researchers, have embraced this technique as a highly flexible way of making solid predictions without prior assumptions.

The second major advance, developed by Packard, Crutchfield, Farmer, and Shaw in the early 1980s, is called state-space reconstruction. Making use of the idea of attractor sets from chaos theory, this method uses geometric considerations to evaluate patterns of behavior. It makes use of an approach known as "delay-coordinate embedding" to plot data from time series as points in a multidimensional space.

The basic idea is as follows: First, choose a time lag in which you wish to review your data. For example, in examining a temperature time series, you might choose "once per hour" or "once per day" as your time lag. Let's assume you select "one hour" as your time lag.

Next, draw coordinate axes corresponding to "the current time," "one hour later," "two hours later," and so on. (If you are using days or minutes instead of hours, adjust accordingly.) Plot each value of the time series on these axes. For example, plot the temperatures at 1:10 P.M., 2:10 P.M., 3:10 P.M., and so on, as a single point. If these temperature values correspond to 30 degrees, 32 degrees, 24 degrees, and so forth, then plot a point at coordinates (30, 32, 24 . . .). Continue until the entire set of data is exhausted.

Then, observe the qualitative and quantitative results of your plot. Do the data seem to spread out over large portions of available space? Is its dimensionality (calculated through special mathematical methods) very high, comparable to the space in which it is embedded? In such cases, the time series plotted would likely have random origins—radio static, for instance. If, in contrast, the dimensionality seems relatively low and the data seems concentrated within certain well-defined bands or regions—attractor sets—the time series would probably have deterministic causes. A straight line, ellipse, or other simple geometric shape (in multiple dimensions), would indicate regular, periodic behavior. A self-similar figure, on the other hand, a fractal akin to a Lorenz attractor, would indicate the presence of deterministic chaos.

Finally comes the interpretation stage. By analyzing patterns found and comparing them to others, you might be able to unravel some of the causes for such behavior. In some cases, by completing or extending the patterns, you might find it possible to anticipate future developments. All in all, you would gain valuable information about the dynamics of the process you are studying.

A number of medical researchers, such as Gottfried Mayer-Kress of Penn State University, Paul E. Rapp of MCP/Hahnemann University, and others have applied state-space analyses in examining the electrical activity of the brain. Their consensus is that the fractal dimensionality of an EEG plot increases with functionality. Although low dimensions correlate well with abnormal brain conditions, such as epileptic seizures and Cruetzfeld-Jacob disease, high dimensions correspond to healthy, alert states. Although some researchers have found indications that fractal dimensionality grows even higher when a subject is performing tasks of greater complexity—counting backward from 100 by sevens, for example—others have found no such correlation; therefore, research on this question continues. Scientists hope that these methods will eventually prove sophisticated enough to diagnose otherwise undetected brain diseases and anticipate mental disorders (schizophrenia, for instance). More and more—in neurology, cardiology, and other branches—complex systems analysis has become an integral part of medicine.

Candid Chromosomes

Improved predictive techniques in medicine raise important ethical issues. It is one thing for patients to know that they might soon experience conditions treatable through drugs or therapy—but what about cases where they might find out years in advance that they will likely develop an untreatable disorder? In such cases, should physicians let patients know, and perhaps have them suffer lifetimes of anguish? Or should doctors withhold critical data, taking a risk that patients left in the dark might later feel cheated out of vital knowledge? Also, in what cases should information be shared with insurance companies? If a state-space portrait of someone's heart rhythms indicates that he is at high risk for ventrical fibrillation, should this information appear on his chart and perhaps become a matter of public record? What, then, if he loses his insurance coverage as a result?

With the success of the government-funded Human Genome Project and of an independent effort privately conducted by the Celera Genomics Corporation, these questions have become increasingly relevant. The groups' ambitious program to map out and sequence all forty-six human chromosomes has effectively reached completion—five years ahead of schedule. Scientists now know the exact positions of the more than 3 billion letters (DNA bases) in the set of instructions for how to assemble a person. Now researchers are beginning to interpret and apply this data, inaugurating an exciting—but potentially frightening—new era for biomedical science.

Full understanding of this information would make it possible for specialists to identify—and in some cases correct—a wide range of genetic disorders. Already, they have developed tests for many of these. Moreover, enhanced data would allow them to predict many adult attributes of children—at birth or even while still in the womb—such as probable height, likelihood of certain diseases, and so forth. If genetic determinists prove correct, researchers would be able to anticipate aptitudes and weaknesses as well, providing parents with rough estimates of the likely intelligence, strength, and dexterity of their progeny. Taking this to an extreme that borders on absurdity, moms and dads would know years in advance whether to start saving for a piano, an artist's studio, a home gym, or a prestigious university.

Of course, these would all be probabilities—best guesses through statistics. Naturally, as chaos theory tells us, small influences after birth could make a whopping difference in a child's life. No one fully understands the relative impact of genetic versus environmental factors; as we've discussed, the nature/nurture debate still rages. Some researchers even speculate that conditions inside the womb could exert the greatest influence on young beings, affecting future chances of heart attacks, strokes, and other ailments. According to Australian physiologist Miodrag Dodic: "You always hear about exercising, keeping weight down and the genetic component of diseases, but the impact of intrauterine life could be even greater."[6]

Despite the ongoing debate about the effects on genes on destiny, there are certain arenas in which virtually everyone would acknowledge genetic predestination. Consider the case of Huntingdon's disease, an incurable hereditary illness in which a mutation in one gene causes a particular protein to be manufactured improperly by the body. For reasons unknown, the production of this altered protein impedes vital parts of the brain, particularly those concerned with movement. By middle age, those affected by the disease begin to lose motor control. Gradually, they relinquish their abilities to walk, move their hands, eat unaided, and speak. Ultimately, they die—experiencing the same sad fate as folksinger Woody Guthrie.

In the early 1990s, geneticists developed a predictive test for the disease. Since then, those for whom it runs in the family have faced a painful choice. If they refuse the test, they face years of anxiety about whether or not they will suffer a drawn-out, premature death. Moreover, if they decide to have children, they pass onto their offspring their genes and their angst.

If, on the other hand, they do decide to take the test, they often must deal with different dilemmas. If the results are positive, they might well experience depression, perhaps incapacitating them emotionally long before the disease takes hold. As Cicero pointed out, referring to the case of Caesar, perfect foreknowledge without ability to change a bad situation might prove lethal to the soul.

Moreover, what if their friends, their employers, and/or their health insurance companies find out about their fatal illness? Might they then be treated differently than they would be otherwise? If they are about to get married, might their newfound knowledge sour their relationships? Do they then have a right to keep the information a secret— even, perhaps, from their fiancées or spouses?

Today, political institutions face the growing challenge of developing new guidelines for the use of predictive information in medicine. The dilemmas raised by enhanced forecasting techniques, such as those provided by the science of complexity and by improved access to information, such as that furnished by the Human Genome Project, demand thorough examination and systematic response. Otherwise we might find ourselves in Cassandra's position—deluged with knowledge but unable to react effectively.

Having developed detailed techniques for biological and medical forecasts, a number of complex systems researchers, including several from the Santa Fe Institute, have refocused their efforts toward a different kind of prediction: anticipating social, technological, and economic trends. Like human bodies, societies might be healthy or ill, flourishing or stagnant. Their overall behavior emerges through the interactions of numerous components. By studying these dynamics through powerful mathematical models, investigators hope to catch a glimpse of the world to come.

7

THE ROAR OF TOMORROW

Social and Technological Forecasting

There is a tide in the affairs of men,
Which, taken at the flood, leads on to fortune

—WILLIAM SHAKESPEARE
(Julius Caesar)

The Prediction Industry

In 1991, hankering for a new challenge, Doyne Farmer left his position at Los Alamos and began to apply his understanding of prediction to a novel domain: the complex world of the stock market. Joining him was his old buddy Norman Packard, similarly self-liberated from his post at the University of Illinois. Together they founded The Prediction Company, an innovative firm devoted to mapping out market patterns and applying scientific and statistical information to investment.

"If I spent my whole life at Los Alamos," Farmer later reflected, "I might have looked up one day and wonder what had happened."[1]

During his decade-long tenure at the Center for Nonlinear Studies, Farmer had kept his sandy hair long—sometimes tied up in a ponytail—a symbol of his hippie roots in an establishment funded by the decidedly unhip Department of Defense. Once he entered the world of high finance, however, he took the momentous step of trimming back his shaggy mane. In time, he traded in his motorcycle for a bicycle, accumulated a wife and three kids, and brought them along for many family-style backpacking trips through Europe. A far cry, it would seem, from his previous days of rambling and gambling.

On second thought, maybe not. According to Farmer, "The market can be viewed as a pure anticipatory game. The market maker plays the role of the casino. Each player attempts to forecast the aggregate action of the other players and bets accordingly. Players that make accurate forecasts are rewarded and those that make poor forecasts are penalized. The average player tends to lose money to the market. However, if a player is good enough, under some circumstances it may be possible to anticipate the other players well enough to overcome the 'house edge' and make a profit."[2]

In establishing their firm, Farmer and Packard tapped into the investment community's omnipresent desire for forecasting expertise. No investor might anticipate every upsurge and downturn in a volatile market. Armed with superior predictive models, however, he might hope to do much better, over time, than standard financial indices such as the Dow Jones Industrial Average. To obtain access to such tools he is willing to pay, and pay dearly.

As forecasting expert Nicholas Rescher relates, "Knowledge is power. . . . It is only natural that a reliance on predictive expertise—actual or presumed—has been ever-present with those responsible for the conduct of large enterprises."[3]

The Prediction Company relies on a variety of mathematical methods to perform its task. Through an exclusive arrangement with a Swiss bank, it receives a certain amount of capital, which its computers, following special strategies, automatically invest. Programs designed by Farmer, Packard, and others review a few thousand stocks every minute, deciding whether to buy, sell, or hold steady. These choices are based on a mixture of classic statistical techniques, machine learning (neural network) models, and "insights from being in the trenches."[4]

Though the firm's directors view markets as evolving, self-organizing systems, they have so far found little use for the standard methods of chaos theory in designing their models. Considering their background—members of the "chaos cabal"—this is surprising. As Farmer explains, though he maintains a strong interest in evaluating strange attractors and related phenomena, "the dimensionality of the market is sufficiently high that these are not useful."[5]

Other complex systems theorists have recently joined Farmer and Packard in applying their skills to the world of business. The Swarm Corporation, established by Chris Langton (along with Glen Ropella and Douglas Orr), the Bios Group, founded by Stuart Kauffman, and Complexica, directed by John Casti (professor of Operations Research and System Theory at the Technical University of Vienna and member of the external faculty of the Santa Fe Institute) represent similar joint ventures between scientific and corporate realms. Rather than acting directly as investors, however, these newer firms play the role of consultants—applying predictive models to a vast panorama of business-related tasks, from optimizing inventory arrangements to assessing consumer preferences. Each of these ventures has its headquarters in Santa Fe, New Mexico—which has virtually become the Silicon Valley of business forecasting.

The tools these groups use derive from many of the standard methods of time series analysis—and then some. Regression techniques, neural network learning models, genetic algorithms, and simulated annealing are supplemented by special designer-made optimization programs. For example, the Swarm Corporation relies in part on artificial life simulations developed by Langton. Bios Group applies principles of self-organization and emergence to issues of management and decentralized control. Finally, Complexica uses, among its methods, a simulation program developed by Casti that examines the risks to companies of natural catastrophes such as earthquakes, hurricanes, and economic collapse. Each group focuses the sharpest predictive lenses available on the branch of human commerce it studies, offering clients glimpses of how their various pursuits might function most efficiently.

War of Anticipation

Modeling human behavior through prognosticative schemes is not just a recent endeavor. Modern social scientific forecasting dates back to the invention of the electronic computer around the time of World War II. Hitler's abominable actions demanded vigilant response, using the best military intelligence technologies available. In a race for the future of humanity, the Allies could ill afford to overlook any possible sources of predictive information.

Then, with the launching of atomic bombs over Hiroshima and Nagasaki, the hot war ended, but a far longer Cold War began. Newly designed computers, vacuum-tubed titans the size of rooms, crunched reams of collected reconnaissance data—dispassionately selecting targets like men throwing darts.

The nuclear arms race quickly evolved into a test of nerves. As satirized in Kubrick's *Dr. Strangelove,* American and Russian scientists competed with each other to develop weaponry of increasing destructive power. Soon, both sides realized that full-scale nuclear war would likely represent the end of civilization itself. Still, neither force would back down, preferring instead to exploit the threat of its arsenal as a political tool. Warfare had become a game of anticipation, and strategic computer modeling was considered a possible route to victory.

The foundation of the RAND Corporation in 1948 represented a strong commitment by the U.S. military toward enhancing the science of forecasting. As the first predictive "think tank" it attracted, over the years, many of the brightest minds of the field, including Herman Kahn, Olaf Helmer, Nicholas Reschler, and (in the 1970s) John Casti. Though military planning constituted its initial *raison d'etre,* it gradually encompassed social and technological prediction as well. Resonating with a growing popular interest in the science of the future, it even helped spawn a societal movement: futurology (also known as futurism).

Though the quest for future knowledge dates back to the dawn of history, futurology as a field is little more than half a century old, arising only after science seemed to have furnished means for addressing that passion. Computer modeling appeared to allow the tracking of social trends toward their most likely outcomes. Perhaps, many in the mid-twentieth century believed, not just the arms race but the human race could be probed by use of properly wired mechanical marvels.

Isaac Asimov's fictional *Foundation* saga, written in the 1940s, captured well the spirit of a future-oriented era. Asimov pondered a mathematics of human behavior so advanced that it could anticipate the triumphs and tragedies of history. Psychohistory, as he dubbed this imaginary science, could predict "the reactions of human conglomerates to fixed social and economic stimuli." However, he cautioned in his tale, "the human conglomerate [must be] sufficiently large for valid statistical treatment . . . [and] unaware of psychohistoric analysis in order that its reactions be truly random . . . "[6] If such conditions were met, he wrote, then the chronicles of millennia to come would unravel like an opened parchment.

Asimov wrote his far-reaching tale decades before the lessons of chaos theory delineated the limits of Laplacean determinism. Even in those more audacious times, he was careful, nonetheless, to state that individual behavior could not be anticipated, and that unforeseen events might disrupt all attempts at planning. Worlds, not people, could be plotted. (In the 1980s, in sequels to the *Foundation* saga, he backed off from his deterministic hypothesis even further, postulating groups with mind-controlling abilities, fine-tuning the original historical scheme.)

The Collective Wisdom of Delphi

Computational models of the 1940s and 1950s were a far cry from Asimov's vision. They could hardly predict the next week's weather, let alone anticipate nuances of human behavior. Yet in the bold skyscape of those luminous days, unclouded by the limits posed by chaos theory, visionaries sought the sixth sense of the oracles and astrologers—albeit through science, not magic.

Suddenly, science fiction seemed less like trifling adolescent fantasy and more in the realm of thoughtful models of the future. The borders between fun, speculative storytelling, and serious, far-reaching research continued to blur (ultimately leading to mixed-source phenomena such as scholarly articles about time travel). When Arthur C. Clarke designed a communications satellite and then predicted, in his tales, an age of long-range telecommunications, the science fiction writer as sage was born.

Indeed, one of the two major predictive approaches developed around that time (the 1950s) at the RAND Corporation borrowed some of the sense of science fiction. The "scenario method" of futurology, proposed by Herman Kahn and refined by his followers, pondered the possibilities of the future through carefully developed scripts. Potential crises and opportunities would be dramatized on paper, as if they were scenes from a political play. The actors would be the leading political, military, and industrial figures of the times. Whenever alternatives existed, they would be played out on the stage of hypotheses as well, until their logical consequences would manifest themselves. Hence, like fine Broadway productions, each script would present growing tensions between realistically drawn characters, an inevitable climax (warfare), and then a natural denouement.

Kahn's instrumental text, *On Thermonuclear War*, published in 1960, introduced his approach to an eager audience. At the height of the Cold War, when teachers drilled their pupils regularly on how to respond to attacks and bomb shelters formed a universal feature of public buildings, terrified citizens wondered how and when the seemingly inevitable global inferno would begin. Kahn ran through a number of possible scenarios, assigning likelihood to each. By realistically detailing the threat of nuclear war, his book warned politicians to turn back before it was too late.

Kahn's scenario method clearly has its limitations. What if several possible scripts seem equally compelling? Who is to arbitrate disagreements among futurologists? How might individual forecasters work with others to hone and refine their models? A second futurological approach, also developed at RAND, attempts to resolve some of these questions. Named the "Delphi method" (after the oracle) by its creators Olaf Helmer and Norman Dalkey, it provides a means of establishing group consensus in fashioning well-constructed predictions. It is especially useful in cases when subjective opinion carries greater weight than objective evidence—when data is sparse, for example. Though first applied mainly to military questions—particularly to estimating probable effects of massive bombing of the United States—it has proven an important tool for all manner of outcome resolution.

The conventional Delphi approach, proposed in 1953, involves repeated anonymous polling. A moderator presents experts with a series of questions about a particular issue. Either using a rating scale or another form of quantitative response, they furnish their opinions. The moderator collects the questionnaires, then revises the questions according to the answers received. He polls the experts again, continuing the process until he concludes that the experts have reached a consensus.

Consider, for instance, a Delphi designed to establish the best route for peace in a particular region. It might consist of questions about the needs of various countries in that area, their likely responses to certain terms of treaties, and so forth. Experts might be asked to rate which potential outcomes would be most likely to satisfy all parties, and which would be least likely. Based on these answers, the moderator might rephrase the queries, asking questions demanding more specific comments. For example, if according to the first questionnaire most participants felt that Country X ought to give Country Y more land, the second questionnaire might inquire how such a territorial exchange might best be carried out.

Sometimes an urgent problem demands especially speedy response. In that case, participants might elect to engage in what is called a "real-time Delphi." In contrast to the standard version, the real-time alternative involves bringing together those involved—whether live or by means of videoconferencing—for direct discussions. For example, a corporation's real-time Delphi might help it to decide which new products to develop, or which divisions to jettison. By the end of the meeting, if it is successful, the group summarizes its conclusions and prepares, if necessary, to take appropriate action.

Naturally, the Delphi method is limited by the ranges of expertise possessed by those involved in the process. A group of ill-informed individuals, no matter how often it debates an issue, is probably not going to make ideal decisions. On the other hand, a single genius might well develop the best way to solve certain problems. Therefore, whether to follow the Delphi approach must be considered on a case-by-case basis.

New York World's Fair

In ancient times, civilizations constructed vast monuments to their faiths, reflecting their deepest yearnings for the future. The lofty pyramids of Mexico and Egypt, the glorious Parthenon in Greece, the majestic temple of Angkor Wat in Cambodia, the labyrinthine city of Machu Picchu in Peru, and other historic wonders each represented an urge to transcend the boundaries of time and fashion an architecture of the eternal. Though many of these structures now lie at least partially in ruin, the vision and zeal of their designers still permeate their presence.

Modern emblems, such as the Unisphere in Flushing Meadows Park, New York, convey similar aspirations for timeless creation. A gargantuan steel globe now surrounded by rusty rockets and faded facades, it once symbolized public excitement for an emerging technological future at the time of the New York World's Fair (1964–1965). It is hard to think of a more representative icon of the futurology movement.

I visited the fair when I was four years old, and count a flag with a picture of the Unisphere as my oldest souvenir. The high-tech Expo opened my young eyes to an age of promise, far removed from the static-burdened black-and-white televisions and clumsy rotary phones with which I was familiar. As much as I could at that age, I marveled at Picturephones, automated houses, and other newfangled technologies.

The 1960s fair was the second to occupy that site. The first New York World's Fair, symbolized by geometric structures known as the Trylon and Perisphere, opened in 1939 and closed in 1940. Coming right before the war, it could not shake the shroud of gloom that had enveloped a world of spreading fascism. The most memorable exhibit, according to many reports, was "Futurama," an intricate model of the city of the future, sponsored by General Motors.

At the 1964 World's Fair, General Motors updated its famous exhibit with a new space-age vision of the future. The revised Futurama featured several imagined alternative living environments of tomorrow, including an underwater vacation resort reachable only by atomic submarines, a lunar colony with buglike vehicles for expeditions, a year-round commercial port excavated from the ice of Antarctica, and a desert farm irrigated with desalinated seawater.

Futuristic housing formed the theme of several other exhibits as well. A "Space City" displayed in the Ford pavilion depicted computerized vehicles whizzing through metallic corridors in structures far from Earth. A completely underground three-bedroom home situated elsewhere at the fair seemed the safest place to be in the event of nuclear war. Surrounded by a concrete shell, it offered protection from fireball blasts and radiation fallout. Fortunately, those visiting the house were escaping from the summer heat and from the noise of the crowd, not from such a disaster.

The year after the fair closed, an eager core of futurists founded the World Future Society (WFS) as "a nonprofit educational and scientific organization for people interested in how social and technological developments are shaping the future."[7] Publisher of a leading periodical, *The Futurist,* it has served since then as a clearinghouse for a broad spectrum of projections about the ways the world might develop in the times ahead. Each year it releases a report of its favorite forecasts. Recent predictions include the disappearance of retail stores because of electronic shopping, and the production of more seafood by means of aquaculture (regulated environments mimicking ocean conditions). Notable members of the WFS governing board include writer Arthur C. Clarke and former U.S. Secretary of Defense Robert McNamara.

Around the same time as the foundation of the WFS, a group of researchers, led by Olaf Helmer, established the Institute for the Future as a think tank for ideas on how to predict and influence possible coming events. Currently based in Menlo Park, California, it has continued to forecast technological, demographic, and commercial trends, offering numerous predictive services for its clients. Some of the areas it has investigated include health care issues and corporate strategic planning.

In the tradition of visionary writers, such as Edward Bellamy, H. G. Wells, Aldous Huxley, and others, futurists have pondered the possibilities and perils of the world of tomorrow. These modern scientific seers have aspired to emulate the predictive successes of their predecessors—Bellamy's vision (in 1888) of the widespread use of credit cards, Wells's premonition (in 1914) of atomic bombs dropped from airplanes, Huxley's depiction (in 1932) of in vitro fertilization and genetic engineering. Bellamy set his sights on 2000, a year that has been, until its arrival, a special target of futurists' gaze.

The Year 2000 as It Could Have Been

The Commission on the Year 2000, a group of thirty researchers selected in 1965 by the American Academy of Arts and Sciences to speculate what the millennial year would be like, arguably constituted the capstone of futurism. Chaired by Daniel Bell, it shuffled a multifaced deck of alternative scenarios, pondering which hands would most likely be dealt. The possibilities it considered ranged from "no-brainer" extensions of the capabilities of those times (computers will be faster; nuclear power will become more widespread), to bizarre modifications without precedent (humans might learn to breathe fluids; the solar system might be artificially rearranged). As in Everett's Many-Worlds model and Deutsch's multiverse approach, no conceivable option, within the bounds of reality's laws, went unexplored. Some of its predictions were remarkably prescient—the everyday use of personal pagers, for example. Others (which the group conceded were radical) hit far from the mark—immortality, telepathy, antigravity, and additional possibilities lifted from the pages of science fiction stories.

Herman Kahn and colleague Anthony Wiener (both members of the Kahn-founded Hudson Institute) summarized the commission's findings in a 1967 book, *The Year 2000: A Framework for Speculation on the Next Thirty-Three Years.* As denizens of the era which it describes, it is illustrative to look back and analyze the scenarios foretold. The report is conveniently arranged into sections on scientific and technological predictions, as well as economic, social, and political forecasts. In addition to what it calls "standard scenarios" it also provides a catalog of alternatives deemed less likely.

Remarkably, even though Kahn and Wiener's book tries to cover all bases, one finds little of today's world encapsulated within its pages. Few of its political projections have turned out to be accurate. For example, it wrongly states:

"Germany is likely to remain divided."[8]

True, practically nobody foresaw the fall of the Berlin Wall. But consider its projections about India's and Japan's respective positions on nuclear weapons:

"There is a good chance that India will abstain." [9]

"There are reasons for conjecturing that Japan might become nuclear power number six."[10]

In fact, while India has openly tested nuclear weaponry, Japan has remained proudly nonnuclear in its approach. So have Sweden and Switzerland, supposed by the authors to have secret interests in joining the nuclear club. Moreover, the book's projection that "by the 1990's as many as fifty countries might have access to such weapons,"[11] has fortunately proven false as well.

The book also incorrectly projects that the Soviet Union will remain intact as an authoritarian superpower, that China and Taiwan will be recognized by the United Nations as two independent states, and that an independent East Germany will be one of the ten richest countries, per capita, in the world. Though it rightly foresees Eastern European independence, it wrongly supposes that the people of the former Soviet bloc would democratically chose to align themselves with their former occupier.

In the sociological realm, *The Year 2000* is even more off the mark. Like many predictive works of its time—Alvin Toffler's *Future Shock* being a prime example—it wrongly imagines the current age facing an abundance of leisure time. According to its projections, Americans should be reveling in four-day workweeks, while planning their annual thirteen-week-long vacations. With retirement taking place earlier and earlier, it supposes, hobbies, sports, and other avocations should have replaced (or at least matched) jobs as the principal focus of people's lives.

Obviously, nothing is further from the truth. In general, people have been working harder and harder. In America, retirements are taking place later and later. In Europe, and in many other parts of the world, the tradition of midday breaks, or siestas, is rapidly disappearing. Shops are open longer and longer hours, run by armies of bleary-eyed, caffeine-fueled workers. Only on certain remote tropical islands occupied by expensive resorts has excessive leisure time remained a burning issue.

Finally, we turn to the book's multitude of scientific and technological forecasts. Here we find both hits and misses. Its cornucopia of medical predictions has turned out to be remarkably apt. The authors forecast widespread use of lasers for surgery, employment of genetic modification techniques, new reliance on computerized medical databases, the introduction of techniques leading to improved heart attack survival and prevention rates, more common use of artificial parts, and continued progress in the reattachment of limbs. Each of these has splendidly come true.

Computers are another area in which Kahn and Wiener largely succeeded in their projections. Though they, and almost no one, anticipated the rise of small personal computers, they came close to having it right. They predicted a computer *console* in every home, remotely connected to large central processors. In this form, computers would find ubiquitous use, from handy library access to paper-free exchanges of currency, and from police criminal checks to student homework assistance. Right, right, right, and right again.

On the other hand, I'm sad to report that no robots clean my household. I've seen very few feature-length 3-D movies, and none of them have been holographic. None of my friends or relatives have been cryonically frozen and revived. I've never traveled on an individual flying platform, nor, except at airports, stood on a moving sidewalk. These are all listed in "One Hundred Technical Innovations Very Likely in the Last Third of the Twentieth Century."

To be fair, I do have a VCR, have often used a FAX machine, and know many people with access to satellite television. These items are also included in the list of predictions. Thus, in summary, though most of the political and social analysis and many of the scientific and technological projections contained within *The Year 2000* have proven inaccurate, a fascinating measure of the predictions regarding medicine, computers, and communications technologies have turned out to be true—a mixed record, for sure.

In *The Popcorn Report,* a bestselling book published in 1991, market forecaster Faith Popcorn similarly tried to anticipate the trends of the current era—with comparable mixed success. Its final chapter lists a number of predictions for the late 1990s. These include:

A quest for knowledge will take over and be strengthened in places
     like Brain Gyms, where we will exercise our thinking . . .
Time release face-lift implants will de-age you ever so gradually.
Fast-food places will offer fast but healthy baby food . . .
We finally recognize the power, insight, and intuition of children
     and turn to them for expert advise, placing them on our most
     important boards, electing them to political office, and making
     them peace arbitrators.
You won't see humans driving buses, at supermarket checkouts or
     serving up fast food. They'll be replaced by colonies of androids who can walk your dog or fight your war.[12]

Popcorn was more on target discussing general societal tendencies of her time, rather than rendering specific predictions about the future. For instance, when she wrote about "cocooning" (the desire to isolate oneself in homey environments) or about "down-aging" (the tendency for older people to act younger), she captured truths apparent then as well as now. Like Kahn, Wiener, and countless other pundits, however, she could not know precisely what social and technological circumstances the years ahead would bring. As she conceded, "No one know knows (exactly) how the future will feel and unfold . . . "[13]

False Calls and Lock-Ins

Why do political and social forecasts often fail? In an age of potent mathematical models and powerful computers, why can't analysts see beyond the current bend of history's stream? What is it about human decision making that in many cases defies all experts? In short, why hasn't futurology been able to do better?

To answer these and related questions, social scientists have pointed to several factors that stem from the practical constraints of current predictive methods as well as from the limits posed by natural boundaries of knowledge. For example, it's an old saw that computer programs and other mathematical algorithms cannot transcend the prejudices of their programmers. Often, the choice of which data points to include and which to leave out, which type of curve to try and fit (linear, exponential, periodic, and so on), what time increments to use (such as years, decades, or centuries), and other aspects of models depend on the instincts and preferences of researchers. Naturally, the scholarly community, with its rigorous peer review, provides safeguards in extreme cases; no one could publish an article in a respected journal in which he developed a long-range prediction based on only one or two data points, for instance. However, in many cases projections based on reasonable suppositions still harbor considerable room for opinion.

The Delphi method is designed to help in cases of dispute. As critics have pointed out, though, it is limited by many factors, including the scope of questions chosen, the knowledge and feelings of participants and even the leanings of moderators (chosen for their impartiality). It is a well-known phenomenon that the answers to surveys often depend on the way questions are phrased.

Consider the following two queries:

"If you found yourself lacking health insurance, because of the high cost of premiums, and were suffering from a long illness, should the government pay for some of the expenses?"

"Should taxpayer dollars go to those delinquent in their insurance payments?"

The same person might vote "yes" to the first query, and "no" to the second, simply because of the phrasing. One social scientist might use the first response to predict expanded government health coverage, another might use the second to forecast a taxpayer's revolt in which government health coverage is eliminated.

In centralized socialist economies, government planning plays a major role in deciding which new products to introduce. The type and quantity of items produced often depends on forecasts for what consumers might need. If these predictions are off, then overstocking or shortages might occur. In capitalist economies, in contrast, the laws of supply and demand presumably steer the market toward providing the most useful merchandise most efficiently. Economists often hail this feature as one of the primary advantages of free enterprise over a planned economy.

But could consumers sometimes make mistakes—wrongly demanding merchandise of inferior quality, while ignoring better products? Might customers latch onto items of interest, became comfortable with them, and then refuse to switch when their quality has long been surpassed? Could free market mechanisms thereby fail if the obsolete remains the standard?

The notion of "lock-ins" finds strong basis in complex systems theory. Given a particular fitness landscape (the "terrain" in graphing space of how good various elements are), certain optimization processes might converge to a medium-size fitness peak (a suboptimal result), rather than the highest (the very best result). That is because the algorithm might not realize that higher peaks exist.

By analogy, imagine a dexterous hermit, determined to live on the highest mountain in the world. His adventures take him to Alaska, where he scales peak after peak, climbing higher and higher. He arrives at Mount McKinley—the tallest mountain in North America—and makes his way to the top. At last, he concludes, he has reached the top of the world. No peak around him seems higher, as far as he can see. Finally content, he remains there until the end of his days, blissfully unaware of the existence of Mount Everest (which stands nearly 50 percent higher above sea level). In essence, he has become "locked-in" to the suboptimal.

Most predictive schemes in the social sciences, particularly in economics, assume that people make choices on the basis of maximizing their own benefit. Cases in which consumers elect to retain the suboptimal are far harder to anticipate. Thus, though nearly everyone at the New York World's Fair hailed the coming of the Picturephone (a telephone with a camera, transmitting images as well as sound), it never caught on—at least partly because of consumer resistance. Because practically nobody had one, nobody wanted one, and the product met a hasty demise. (Until, starting in the 1990s, digital cameras and the Internet made image transmission more popular.)

The QWERTY Quandary

In a 1985 research article, Stanford economist Paul David introduced what he saw as the classic example of a locked-in consumer blunder, namely the current standard typewriter keyboard. Officially known as the Universal keyboard, but more commonly known as "QWERTY" (after the first six letters of its pattern), it has appeared on virtually all typewriters for more than a century. Yet, as David pointed out, it is certainly not the most efficient way of arranging the letters for fastest typing.

The QWERTY keyboard was developed by inventor Christopher Sholes in the 1860s to remedy some of the mechanical problems of early typewriters. Studying the machine's inner mechanisms and noting the frequency of English language letter combinations, he deliberately arranged the keys such that the typebars from common pairs strike the paper from opposite directions. For instance, he placed "e" and "l," a frequent combination, on opposite sides of the keyboard. His motivation was to prevent keys from jamming, a common complaint about the machines of the day.

In 1873, Sholes sold his patent rights to the Remington corporation. Remington used the design to produce millions of typewriters over the years. Because Remington's machine became the standard instrument, the QWERTY keyboard became the standard as well.

Not that QWERTY was the only contender. In 1936, Professor August Dvorak introduced his own design, in which commonly used letters occupied a more central location than in QWERTY. He ran a number of tests, attempting to prove that his keyboard was superior. Indeed, typing tests that he ran, as well as experiments done by the U.S. Navy, seemed to show that the Dvorak system permitted significantly faster typing.

Yet today virtually no keyboards use the Dvorak method. If it is superior, why hasn't it been adopted, particularly in an age in which sticking keys certainly does not matter? David argued that the only reason QWERTY has been retained is because it has become so much the standard that consumers are unwilling to change. Without willing typists, new typewriters won't be built, and without new machines almost nobody would want to learn a new system. Thus, trapped in a vicious cycle, the public has become "locked-in" to an inferior product.

Echoing David, other economists have cited additional examples of what they call "path dependence." Path dependence, a term borrowed loosely from physics, is the tendency of the market to reflect historical factors (for example, decisions that turned out incorrect), as well as current motivations (such as the best product for the cheapest price). Besides QWERTY, other famous cases of path dependence, according to these experts, include VCR format standards (Beta versus VHS) and choice of computer operating systems (Apple's versus IBM's).

Five years after David published his thesis, two other leading economists, S. J. Liebowitz and Stephen E. Margolis, published a sharp rebuttal titled "The Fable of the Keys." Reviewing the record of typing competitions, they argued that Dvorak's claims of clear keyboard superiority were unsubstantiated. Citing bias, they called into question any contests that Dvorak directly supervised, and found little of substance in the record otherwise. In fact, they discovered a number of examples in which QWERTY bested other designs. The QWERTY keyboard, they concluded, has survived not just because of chance circumstances (path dependence), but rather because it is among the best designs. Perhaps it is not *the* best, they wrote, but it is good enough to justify keeping it and avoiding the expense of a changeover.

With battle lines solidly drawn, the QWERTY skirmish has continued to rage among the economics community—pitting those who believe in economic planning to avoid market failure against those who support unrestricted markets. It seems to me that both sides have made valid points. In the days of manual typewriters, the QWERTY scheme had much to recommend for itself. With sticking keyboards a major issue, it's unclear if Dvorak's system would have represented much of an improvement. Today, however, there are computer keyboards much more convenient than QWERTY, with common letters located in easier-to-reach places. Yet because of QWERTY's "lock-in," I doubt that it will be soon be replaced.

The Perils of Punditry

If consumers could anticipate the technologies of tomorrow, they would likely be much more careful in their purchases. If politicians could anticipate the social movements of tomorrow, they probably would be much more cautious in their actions and statements. Nobody wants their current words or deeds to brand them obsolete in coming years.

We have discussed a number of practical reasons why social scientific forecasting is often difficult or impossible. Unlike in the natural sciences, such predictions are generally difficult to quantify. Data and analyses are often subject to the vagaries of human interpretation; polls can similarly be misleading.

Even well-defined, mathematically rigorous predictive approaches, such as genetic algorithms for optimization, sometimes become trapped in suboptimal regimes, yielding solutions that are well off the mark. (Researchers have proposed a number of ways to try to deal with this problem; the details lie beyond the scope of this book.) For example, in trying to anticipate the route that a traveling salesman would take (to cite the classic problem), an algorithm might predict a less-than-ideal path, failing to forecast an even better route. Substitute "government," "corporation," or "popular movement" for "traveling salesman," and "policy" for "route," and one might see some of the shortcomings of even well-posed, quantitative analyses.

As philosopher Karl Popper has pointed out, even if social scientists someday resolve these practical matters, certain fundamental issues about the limitations of future knowledge will always remain. In "The Open Universe: An Argument for Indeterminism," he brilliantly fashions notions from Kantian philosophy, quantum physics, and

Gödelian logic into a cogent argument that the human race cannot predict its own destiny. The main thrust of his reasoning is that because we cannot anticipate the knowledge our descendants will have, we cannot forecast what they will do. Why can't we reason out in advance what they will know? We are preventing from doing so, according to Popper, because since our brains and machines run at least as slow as theirs, it would take so long for us to acquire such information that the future would have already arrived.

In the spirit of Kant, Popper asserts that we cannot step out of the universe and examine its every mechanism. Echoing Gödel, he also points out that a system cannot fully grasp itself. Regardless of the Newtonian laws (as modified by quantum principles) that govern its interactions on the physical level, our complete picture of reality must be based on our own perceptions on the human level; that is, from *within*. And our own intuitive experiences, as mortal denizens of our world, suggest that free will must be part of the equation.

Free will, Popper contends, certainly does not mean that human actions are acausal. Naturally, each effect much have a cause (ultimately, the firing of neurons in the brain). Rather, it means that unless one could enter somebody's mind, one could not predict what the person would do under any given set of circumstances. Even self-prediction, he argues, would involve sifting through the reasoning and accumulating all the data that would impel future actions—a process that would take as long as living through the experiences oneself.

In other words, to use complex systems terminology, human thoughts and actions are computationally indeterminate. As in the case of Chaitlin's numbers and other examples of Turing's Halting Problem, to anticipate the future of someone's mental processes (even those of a future version of oneself), one must experience everything that person does step by step. The prediction takes just as long to unfold as reality.

Let's say we wished to forecast the events of the year 3000. At the very least, we would have to predict the growth of social, political, and technological movements from now until then. This would require anticipating the blossoming of human knowledge and experience during the next thousand years. Moreover, it would entail understanding how that information would be interpreted and reinterpreted by billions of people in centuries to come. Finally, it would involve mapping which

trends would be most likely to amplify themselves, which would probably continue as they are, and which would presumably die out. The
butterfly effect, in which a small change might make a whopping difference, would render such a predictive process all but impossible.

For these reasons, it's interesting to note that most science fiction visions of future human behavior tell us more about the world in which
the story was written than they do about the world of tomorrow. Consider, for instance, the original *Star Trek* television series, produced in
the 1960s. Though its theme was life in space in centuries to come,
many of its episodes reflect the social movements of the Vietnam War
era (hawks vs. doves, the hippies, and so forth). It would be surprising
indeed if human life in the future resembles its far-flung depiction.

Does that mean that nothing can be predicted years in advance?
Hardly. As any scientist would attest, myriads of successful predictions
are rendered all the time. Typically, though, these accurate forecasts
have nothing to do with long-term human behavior, and everything to
do with well-understood physical mechanisms. Despite the restrictions
on knowledge posed by chaos theory and quantum uncertainty, large
sectors of nature do operate, for the most part, like clockwork. Otherwise, Kepler wouldn't have been able to discern the laws of planetary
behavior, Newton couldn't have unlocked the principles of gravitation,
Bohr couldn't have deduced the energy levels of atomic electrons, and
science as we know it would not exist. Often, in the cases of systems
with large numbers of components, when direct modeling is impossible, statistical laws kick in. Thus, in the gambling halls of long-term
forecasting, mechanistic natural systems, particularly in astronomy
and physics, generally offer far better bets than do ever-changing human projects and organizations.

8

TIME'S TERMINUS

Visions of the Far Future

I strained my fainting intelligence to capture
something of the form of the ultimate cosmos.
With mingled admiration and protest I
haltingly glimpsed the final subtleties of world
and flesh and spirit . . .

—OLAF STAPLEDON
(Star Maker)

Billions and Billions

In the far, far future, when Earth and its inhabitants are long gone and the Sun has faded from the sky, the cosmos will be a much bleaker place. According to astrophysicist Michael Turner of the University of Chicago, 400 billion or 500 billion years from now space will be so dilute that from any vantage point one will be able to view only a handful of nearby galaxies. The rest will have either died out or moved so far away that observation would be impossible.

Theorists Fred Adams of the University of Michigan and Greg Laughlin of the University of California at Berkeley agree with Turner's assessment, and project even further into the eons ahead. They have aspired to chronicle the next 10^{100} (one followed by 100 zeroes) years of cosmic history and beyond, detailing the tortoise steps the universe will take as it approaches its ultimate demise.

Imagine the reactions of economic or social forecasters to such extraordinarily long-range predictions. No doubt many would find such multitrillion-year timetables incredibly audacious. Considering the difficulties social scientists often have in predicting what will happen decades in the future, they might wonder how cosmologists muster the confidence to address far more distant eras. How could Carl Sagan, for instance, speak of ages "billions and billions" hence, when Herman Kahn and his contemporaries in the 1960s couldn't even anticipate the rise of the Internet in the 1990s?

There are several key differences between cosmology and other fields that lend a lot of chutzpah to those who study the future of the universe. First and foremost, unlike many other areas, cosmology is based on exact, deterministic scientific equations: Einstein's law of general relativity. As Eddington's 1919 discovery and countless other experiments have confirmed, Einstein's gravitational principles successfully model the observed behavior of space, time, and light. Some theorists have proposed small variations in these equations to accommodate certain unexplained phenomena (forces exerted by hidden matter, for example). Typically these involve the addition of one or more extra terms. Other researchers maintain that general relativity is fine as stands. Whether somewhat modified or maintained as they are, the set of principles discovered by Einstein offer cosmology substantial forecasting power.

Another advantage possessed by cosmological prediction is an ancient, visible database: the sky. In climatology, seismography, economics, and almost every other prognosticative field, experts consider themselves lucky if they can consult detailed records more than a few decades old. Moreover, they are disinclined to render forecasts that extend known data much beyond the amount already known. No decent analyst, for instance, would predict twenty years ahead if his records of past behavior date only five years back. In cosmology, on the other hand, the heavens provide a visual account of what happened billions of years ago. Light dating as far back as the dawn of cosmic history continually rains down on Earth—permitting astronomers ready access to age-old information about the universe. As astronomical instruments continuously improve, they have been able to observe and analyze more and more information about the universal past. Because they can see back so far, they feel freer to project so far ahead.

Key philosophical postulates offer cosmology additional benefit. The cosmological principle, a common-sense statement that Earth occupies an average place in the universe, serves surprisingly well to rule out many astrophysical theories. This assertion constrains cosmology to consider only homogeneous (looking identical from all vantage points) solutions of Einstein's equations of general relativity. If we believe that we aren't penthouse dwellers, but rather typical tenants of the cosmic apartment building, we must assume that we have about the same view as everyone else. Therefore, from any position, space should look pretty similar.

Another conjecture, the anthropic principle, is somewhat more controversial. Of all possible universes, it states, we live in one that possesses all the factors to support intelligent life; otherwise we wouldn't be here to observe it the way it is. No one disputes this statement; the controversy stems from its interpretation. Some cosmologists—Frank Tipler, for instance—contend the anthropic principle limits the range of viable universes to only a tiny fraction of the set of all possibilities. According to this view, our cosmos is extremely special. If it was just a bit different, we wouldn't be here. Other researchers—Roger Penrose, to name one—dispute this contention. They argue instead that the parameters of the cosmos (the gravitational constant and other physical measures) are not so fine tuned; minor differences would have still led to the world as we know it.

One remarkable feature about the present-day cosmos is its apparent isotropy. Isotropy, in this context, means appearing the same in all directions from Earth. No matter which way astronomers point their telescopes they observe similar average distributions of galaxies. True, on certain scales, the cosmos has structure. Stars are arranged into elliptical, spiral, and other types of galaxies. Galaxies, in turn, are grouped into clusters, superclusters, and even greater arrangements. Eventually, though, this hierarchy ends. On the largest scale of observation, space looks eminently uniform.

Assumptions of homogeneity and isotropy sharply limit the range of appropriate solutions to Einstein's general relativistic equations. They mandate that space, as a whole, must have one of three possible kinds of geometries. Like the localized spatial warping represented by black holes, these can best be visualized in higher dimensions.

The first possibility is that the universe resembles a hypersphere: the four-dimensional equivalent of an ordinary ball. If space were to have such a structure, it would curve back on itself in all directions. Any traveler venturing far enough along any given straight line route would ultimately circumnavigate the cosmos and arrive back at his starting point.

The second option is that space constitutes a hyperboloid: the four-dimensional analogue of a saddle. If the cosmos were shaped in such a manner, it would arch away from itself in all directions. Unlike the hypersphere, it would never connect up with itself. Infinite in extent, circumnavigation would be impossible.

The third alternative is that space is completely flat: the four-dimensional equivalent of an infinite plane. A flat cosmos would stretch out indefinitely—never curving, never linking up with itself. Consequently, as in the second case but not the first, wanderers could travel forever along straight line paths.

In 1922 Russian mathematician Alexander Friedmann solved the Einstein equations for each of these three geometric possibilities. Friedmann, in his career, had ventured from one area of forecasting to another; he started out in meteorology and ended up in cosmology. In his more famous job, he brilliantly outlined the behavior of the three models he found, thus delineating the isotropic, homogeneous solutions of general relativity.

The three kinds of Friedmann universes are called closed, open, and flat—respectively corresponding to hyperspherical, hyperboloidal, and flat four-dimensional surfaces. Closed universes begin at a point, expand outward until they reach a maximum radius, and then collapse back down to a point. Open models, in contrast, continue to expand forever, never reaching a maximum size. Finally, flat models walk an infinite tightrope, teetering forever on the brink between open and closed.

All three models suggest that the presently observable universe was once much smaller than it is today, and that is has ballooned outward ever since. This hypothesis was confirmed in 1929 by American astronomer Edwin Hubble's monumental discovery that all other galaxies, save those in our immediate neighborhood, are moving away from

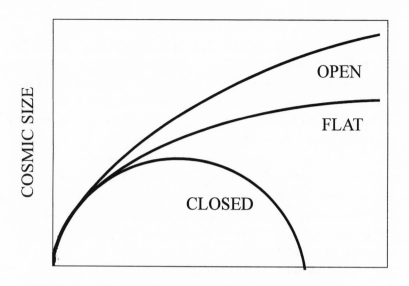

TIME SINCE THE BIG BANG

*FIGURE 8.1   Depicted are three possible cosmic destinies. In the "closed"
scenario, the universe expands for a time, then eventually collapses back
down to a point. In the "open" case, on the other hand, the cosmos continues forever to get bigger and bigger. A third possibility, the "flat" scenario,
represents a universe perpetually teetering on the brink between continued
expansion and recollapse.*

us. Moreover, the farther away from us galaxies are, the faster they appear to recede.

Hubble discovered this principle—known as Hubble's law—by
comparing the distances of galaxies to their recessional velocities.
Galactic distances can be found by "standard candles": objects within
them that exhibit known absolute brightness (rates of energy production). The example Hubble used was a type of star known as a Cepheid
variable. By comparing such a body's apparent brightness (how
brightly it shines, as measured by earthly telescopes) with its absolute
brightness, astronomers can determine its distance—and hence the
distance of the galaxy in which it is located. The effect is similar to the
principle sailors use in estimating the proximity of a lighthouse beacon
shining in the dark; the greater the brilliance, the closer it is.

To ascertain the recessional velocities of galaxies, Hubble and his colleagues employed what is called the Doppler method. In the Doppler effect, the spectral lines of a light source alter predictably whenever the source moves toward or away from an observer. Just as a police car's siren wails at a higher pitch whenever it is approaching, and at a lower pitch whenever it is moving away, light from radiant objects shifts toward higher, bluer frequencies whenever they come closer, and lower, redder frequencies whenever they recede. Spectral expert Vesto Slipher had discovered a number of years earlier that distant galaxies exhibit this effect as well—their light shifting toward the red. When Hubble matched these red shifts to distances, he found unmistakable evidence that the farther away a galaxy, the greater its light appeared to shift, and thus the faster it seemed to be moving away. He thereby concluded that space was expanding.

From the 1940s until the 1960s the astronomical community interpreted Hubble's law in two competing ways: the Big Bang and Steady State approaches. According to the Big Bang hypothesis, championed by physicist George Gamow and others, the cosmos was once extraordinarily hot and compact. All of its matter and energy occupied a ball of pointlike dimensions. Gradually, everything expanded and cooled, coalescing into the celestial bodies we see today.

In the Steady State alternative, proposed by astronomers Fred Hoyle, Thomas Gold, and Herman Bondi, although space has expanded, new objects have continuously filled in the gaps. Consequently, the cosmos has never been hot and dense. Rather, it has maintained the same overall appearance for eternity.

In the minds of most astronomers, the question of which of these models is correct was resolved by the 1965 discovery of the cosmic microwave background. As Bell Laboratory scientists Arno Penzias and Robert Wilson found in their Nobel Prize–winning work, the cosmos is filled with radiation left over from the Big Bang. Through sheer serendipity, they detected this residual energy while tuning a horn antenna in Holmdel, New Jersey, for radio astronomy experiments. To their befuddlement, no matter how much they adjusted the instrument, it continued to pick up a strange background hiss. Eventually, after ruling out other possibilities, they realized that the noise originated from deep space. Because of its marked uniformity—maintain-

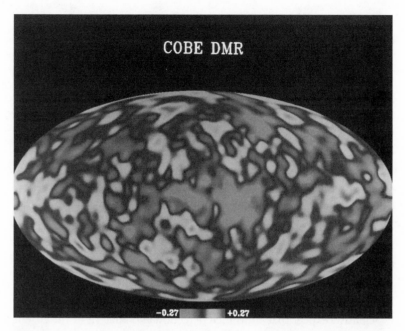

FIGURE 8.2   Depicted is the sky map of the cosmic microwave radiation produced by the Big Bang, as detected by the COBE (Cosmic Background Explorer) satellite. The various shades correspond to slightly different temperatures of the radiation. By analyzing such portraits of spatial conditions, astrophysicists hope to understand the composition and evolution of the early universe. Ultimately, they hope that this data will help unlock the secrets of the cosmic future as well (courtesy of NASA).

ing virtually the same character no matter which way they directed the antenna—they deduced that it was universal in nature.

They measured the radiation's temperature to be approximately three degrees above absolute zero. This matched the value that Gamow had predicted for the Big Bang's radiant energy, cooled down over billions of years. According to Gamow's theory, the heat content of the universe was originally thousands of degrees. In the way that a balloon's hot air cools as it expands, the temperature of space dropped more and more as the cosmos got bigger and bigger. Eventually the cosmic background reached its current frigid level. With Penzias and Wilson's determination that the radiation they found had the temperature anticipated by Gamow, they provided indisputable confirmation of the Big Bang theory.

In the early 1990s, a specially designed NASA satellite called COBE (COsmic Background Explorer) began the process of mapping more precisely the Big Bang's relic radiation. Led by George Smoot of the University of California at Berkeley, a team of researchers analyzed the satellite's data and confirmed the background radiation's essentially isotropic distribution—indicating that it stemmed from the universe itself, not from any particular source. But, as galactic formation models predicted, the radiation's measured distribution wasn't entirely smooth. In a further triumph for the Big Bang theory, Smoot's team discovered minute anisotropies—temperature changes of approximately six parts per million—indicating that the early universe contained the tiny seeds of what later became galaxies. Presented with such impressive results—a wealth of data for theoretical analysis—cosmology gained new prognosticative power.

Scientists expect that continued detection and analysis of the cosmic background radiation will help them further understand the primordial structure of the universe. They also hope that it will allow them to gauge its overall behavior, answering fundamental questions about its origins and destiny. One of the most important issues astronomers hope to resolve is whether the cosmos is open, closed, or flat. Will space expand forever, they wonder, or will it someday contract? Thanks to modern techniques, astronomy may soon unravel these age-old riddles.

Measuring Cosmic Destiny

Sometimes unlocking a bolted door reveals others even more tightly sealed. In the 1990s, astronomy witnessed the boldest attempt ever to gauge the fate of space. The effort produced impressive results—insight into the fundamental dynamics of the cosmos. Yet strangely enough, it raised as many questions as it yielded answers.

In a cleverly conceived experiment designed to gauge how fast the universe was expanding billions of years ago and thereby predict how quickly it will spread out in the future, two teams of astronomers measured the properties of supernovae (stellar explosions) in ancient, extremely remote galaxies. Each group, one led by Saul Perlmutter of the

University of California at Berkeley, and the other by Brian Schmidt of Mount Stromlo and Siding Springs Observatories in Australia, hoped that a systematic account of the distances and velocities of these starbursts would help theorists assess the dynamics of space billions of years ago when the explosions took place. Specifically, how has Hubble's proportionality constant (the rate by which space is expanding) changed over the eons? By methodically locating a number of these supernova, the teams found that they could use each object's energy output to assess how far away it was, and its Doppler shift to gauge how fast its host galaxy was moving away from our own. Finally, they were able to use their data to compute the rate by which Hubble's constant has changed over the past 10 billion years.

Perlmutter describes the reason the groups used distant starbursts as measuring devices:

> Supernovae make great tools for studying cosmology. This is for two reasons: first, they are bright enough to be seen at great distances, so the light has been travelling to us for a significant fraction of the age of the universe. This allows us to compare the early universe to the current universe.
>
> Second, one type of supernova—Type Ia—makes an excellent "standard candle." They almost all reach the same intrinsic brightness, and the small differences in this peak brightness can be determined from the rate with which they brighten and fade. This means that their apparent brightness (as we measure at the telescope) can tell us how far away they are."[1]

The two teams' results were quite unexpected. To their astonishment, both groups measured the distant supernovae to be significantly fainter than theory had predicted. After ruling out mundane reasons for such dimness—intervening cosmic dust, for example—they concluded that either space is curved in a different way than once thought, or that the supernovae they detected are farther away than their Doppler shifts would indicate.

Let's consider these two possibilities. First, examine how spatial curvature might disguise the true radiance of distant stellar explosions. For certain theoretical reasons, related to a conjectured period of the early universe called the inflationary era, many cosmologists have liked

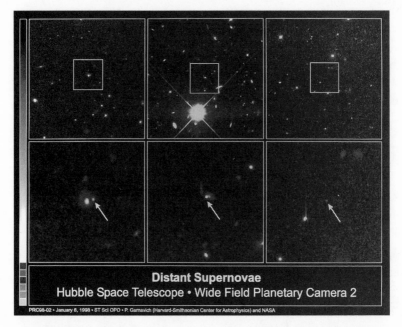

FIGURE 8.3   *These Hubble Space Telescope images depict three extremely distant supernovae—stellar explosions that took place billions of years ago. Astronomers are using such images to pin down changes in the expansion rate of the universe. By understanding the past behavior of the cosmos, they hope to predict its long-term future (courtesy of NASA).*

to believe that space currently resembles a flat Friedmann model. Through a flat space, light travels undistorted. In contrast, if the universe were an open Friedmann model instead, shaped like a hyperboloid, then light would tend to curve as it moved. The radiant energy of a remote supernova would spread out over time, appearing fainter to terrestrial astronomers than it would otherwise. Thus an open universe would help explain the apparent dimness of the supernovae.

An even stronger possibility suggested by the data is that Hubble's law has changed over time. Perhaps the universe expanded far more slowly in the past (that is, with a smaller Hubble's constant). In that case, ancient light emanating from a remote supernova would have less Doppler shift than one would expect for that distance. Consequently, based on the present-day value of Hubble's constant, astronomers

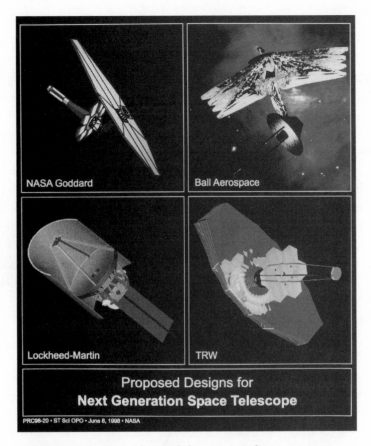

NASA Goddard

Ball Aerospace

Lockheed-Martin

TRW

Proposed Designs for
Next Generation Space Telescope

PRC98-20 • ST ScI OPO • June 8, 1998 • NASA

FIGURE 8.4   To understand the universe's future, we must peer further into its past, using more and more powerful optical instruments. Eventually, we must launch a telescope into orbit that outperforms the Hubble. Here are four possible designs for the next generation of space telescopes (courtesy of NASA).

would mistake the explosion to be closer. With less brilliance than expected for such proximity, they would also perceive it as dim.

If indeed the growth of space was much smaller in the past, then it must have sped up considerably. Extrapolating the value of this acceleration forward in time, theorists project that the universe will expand faster and faster for at least the immediate future. This contrasts sharply with theoretical predictions, based on the Friedmann models, that cosmic expansion should currently be stabilized or slowing down, not speeding up.

Though most cosmologists still assume that general relativity is wholly correct and that space can therefore be modeled according to the simple pictures presented by Friedmann, the findings of Perlmutter, Schmidt, and others have burst open a hornet's nest of dissent. The fact that the universe seems to be flying apart at an ever-increasing rate has driven a number of theorists to propose alternative approaches.

Princeton professor Paul Steinhardt, for example, has postulated the existence of a new kind of matter, called quintessence, that is pushing the cosmos ever outward. Cambridge theorist John Barrow, on the other hand, has demonstrated that the current cosmic acceleration might well be explained by assuming that the speed of light has changed throughout the eons. Others have revived Einstein's discarded notion of a cosmological constant—a type of universal antigravity—propelling the universe outward through negative pressure. Each of these radical proposals would involve alterations in the basic Big Bang approach embraced for many decades.

Naturally, before making radical changes to Einstein's venerable theory, astronomers will need to test the implications of each of these novel alternatives. They hope that a bevy of telescopic measurements with ground- and space-based instruments and greater analysis of the cosmic background radiation will reveal which, if any, of the new proposals should be adopted. Outreaching even the powerful Hubble Space Telescope, researchers expect that the next generation of space instruments will help provide some of the answers to these profound cosmic riddles.

Unveiling the Final Stages

What if someday astronomers knew enough about the universe to picture not just its past, present, and intermediate future, but its ultimate state as well? Researchers Fred Adams and Greg Laughlin have attempted to do just that. In their extraordinary far-reaching simulations, based on what is currently known about astrophysics, they have endeavored to run stellar, galactic, and other celestial processes forward, bringing them to their natural limits. Though their efforts are extremely ambitious—trying to peer 10^{100} years into the future—they are certainly creative. The fact that researchers might even comment on such distant times reveals the predictive strength of contemporary cosmology.

Although Adams and Laughlin portray the slow demise of the universe, they paint an optimistic picture of times ahead. They offer the remarkable hypothesis that the cosmos could bristle with life—in one form or another—for trillions and trillions of millennia to come. Though much has happened so far, they consider the universe to still be in its infancy, relative to long, vibrant years in store.

Like a spectacular theatrical production, full of dramatic set changes, Adams and Laughlin have divided their epic saga of eternity into five distinct acts. The first, which they call the Primordial era, came to a close a long time ago. This interval represents the hottest period of the universe, from its emergence out of nothingness in the Big Bang to the creation of its first stable atoms roughly 1 million years later. Astronomers trace the origins of the cosmic background radiation to the end of this period, when, as electrons and protons assembled into atoms, photons could begin to move freely through space.

The next era, the Stelliferous, comprises the age we live in, when shining stars dominate the sky. Hydrogen-burning suns suckle the surrounding planets with nutritious warmth, fostering the abundant growth of life. On some of these worlds, intelligent beings enjoy the most prosperous period of universal history, transforming free-flowing stellar energy into artful creations. Civilizations rise and fall, but living worlds continue to thrive.

Sadly, even the greatest actors must someday leave the stage. Like the denizens of Earth, stars too are mortal. Once a stellar body's primary source of fuel (its core hydrogen and other fusible material) is exhausted, it will begin its inevitable march toward oblivion. For some, a catastrophic supernova burst will light the funeral pyre. Others will slowly expire like the embers of a campfire. Eventually each will become either a white dwarf, neutron star, or black hole. These burnt-out relics will join large colonies of brown dwarfs: stars that never started shining in the first place. Adams and Laughlin call this dwarf-dominated period the Degenerate era.

The Degenerate era will end, the chroniclers of the far future predict, when matter itself becomes unstable. Protons, along with neutrons, the building blocks of all atomic nuclei, seem eternal. However, many theories unifying the electroweak with the strong force predict that protons must eventually decay. This annihilation, estimated to take 10^{35} years, will release slowly trickling quantities of energy, far less than produced by nuclear fusion, but enough to keep the universe from dying altogether.

Adams and Laughlin speculate that living organisms, even intelligent beings, might survive far into the future if they adapt well enough to extremely low temperatures. By metabolizing slower and slower, like bears in an extreme form of hibernation, they might keep themselves alive throughout the cold, dark days of the end times. Thoughts and movements that take seconds for us would take millions of years for our glacial-age descendants. But if the universe survives as long as Adams and Laughlin suggest, time should present no problem.

Next will be the Black Hole era, when nothing else will exist except the elusive objects that give that period its name. By then, the cosmos will be much larger than today, but will have far fewer objects. Consequently, like a vast battlefield housing only the carcasses of the fallen, it will be an extraordinarily bleak and empty place.

Finally, by the year 10^{100} or so, the black holes will have decayed as well. Entropy will have reached its maximum, and the lifeless state known as "heat death" will ensue. For our universe at least—Adams and Laughlin speculate about successor cosmos—nothing will ever happen again.

Beyond Time's Edge

It's hard to picture death, and it's even harder to picture the death of everything. Though we are mortal, we like to imagine that life itself will always go on. Most religions offer the solace of eternal existence, either through perpetual rebirth of the cosmos (as many Hindus ponder, for instance), or else by means of a final, timeless paradise (as many Christians, Muslims, and Jews imagine). Standard scientific cosmology doesn't offer such options. Once all matter decays and all usable energy is exhausted, all life must cease if it hasn't already.

But what if the scientific idea of an absolute end is too rigid? Applying an Eastern perspective to the possibilities of general relativity, some contemporary theorists look beyond such bleak limitations and imagine new universes emerging from the debris of the old. These visions challenge us to tread the farthest-flung paths of Einstein's theory (in its quantum extension) in a bold expedition to worlds forever hidden from our own. Frustratingly, these speculative notions cannot be independently verified, as the scientific method would demand. Because they pertain to domains beyond conceivable observation, probably no one will ever be able to assess whether they correctly describe physical reality. At least, however, they are theoretically plausible. Their mere possibility offers the tonic of hope for those disturbed by the idea of universal extinction.

Andrei Linde, a Russian-born cosmologist currently at Stanford, has constructed a novel explanation of the origin of the universe that ponders other realms beyond ours. According to Linde, the space around us, as far as the eye (or telescope) can see, has expanded from a simple seed germinated from a vaster cosmos. Other universes, similarly born, lie well outside ours. Although some of these brethren offspring may have galaxies, stars, even life of their own, countless others are still-births: energy formations unable to develop fully.

Imagine a grand cosmic nursery—a hothouse of primordial energy. Nurtured within its soil are myriad scattered seeds, representing randomly generated "vacuum fluctuations." As Heisenberg's uncertainty principle informs us, even the emptiest voids produce small surges of energy that rise out of nothingness, enjoy brief tastes of reality, then disappear. These "vacuum fluctuations" derive their brief existence from the fact that the amount of energy in a particular region of space cannot be known with precise certainty; therefore, it must vary slightly over time.

As Linde has shown, under certain circumstances the presence of an energy field (a vacuum fluctuation, in this case) in a given part of space might induce that region to expand extremely rapidly. Through this process, known as inflation, a tiny seed might sprout into a full-fledged universe. Presumably this happens for a number of seeds within the grand cosmic greenhouse—those induced by auspicious vacuum fluctuations to grow ever bigger. Some of the sprouted universes expand outward more and more, until they are large enough and cool enough to favor galactic formation. One of these became our own familiar domain. Other saplings, on the other hand, never quite make it. They rise up just for a brief period, then collapse back down again. Devoid of life, no intelligent entities within would record such a short-lived space's existence.

According to Linde, this process of germination could repeat itself in the final days of our own universe. In the Black Hole era, space will contain large numbers of ultradense objects. Conceivably, the energetic centers of massive black holes could offer rich soil for new seedling realms. Following Linde's inflationary scheme, some of these could bloom into successor universes. Perhaps a fraction of these could eventually provide living quarters for whole new civilizations. As a consequence, the end times of our own domain could engender the flowering of many others.

Lee Smolin, a gravitational theorist at Penn State University, agrees with Linde that black holes might spawn new universes. In Smolin's more detailed conception, based on Darwinian theory, the cosmos evolves over time by producing numerous offspring that compete with one another for survival. In a grand cosmic struggle, involving countless domains vying for dominance, a single universe might best its competitors by engendering as many successors as possible. As Smolin conjectures, because having a greater number of black holes leads to producing more children, the number of black holes a universe possesses predicts its reproductive success. Coincidentally, he points out, universes with more black holes harbor more of the elements necessary for life. Therefore life-sustaining universes, such as our own, have a greater chance of spawning others.

By analogy, imagine an animal that can only reproduce if it ingests certain nutrients that are only found in fish. Moreover, the more fish it eats, the more offspring it creates. As luck would have it, those animals more likely to enjoy the taste of fish are also genetically predisposed for higher intelligence. Because, as generations pass, the fish lovers will tend to dominate over those that don't like fish as much, the average intelligence of the community will tend to rise. Similarly, because black-hole-filled universes produce more successors, their life-supporting features prevail over time.

The models developed by Linde and Smolin offer the intriguing possibility that intelligence and creativity might transcend the bounds imposed by physical decay. Perhaps the spark of sentient life might somehow survive the quashing of the universe's flame, serving to ignite new fires in successor domains. As Linde speculates, if we could some-how communicate the essence of what is human to intelligent beings in future realms, our legacy might thereby be eternal.

Linde and Smolin recognize that their visions of cosmic destiny transcend the classical scientific paradigm, echoed by Western religious concepts, of reality as a single strand stretching from creation to anni-hilation. Describing their concepts of eternally reproducing cosmos, the theorists draw instead on Eastern spiritual notions of endless reju-venation. In my view, their models exemplify a new concept of time that defines our age and shapes our images of the future.

From neural networks to quantum computation, from genetic algo-rithms to multiple universe models, and from fractal imagery to the World Wide Web, parallelism has supplanted linear determinism as the dominant metaphor of our times. As we begin the twenty-first cen-tury, embryonic tendencies spawned in past decades (starting as far back as the jazz age) have matured into a wholly new way of looking at time and destiny. With complexity pervading so many aspects of life, from the way we learn information to the methods by which we at-tempt to foresee the future, truly we have entered the age of the labyrinth.

CONCLUSION

Frontiers of Prediction

It is not primarily in the present nor in the past that we live. Our life is an activity directed toward what is to come.

—José Ortega y Gasset

The Second Fall

Most ancient cultures harbored a belief that all forms of time run in cycles. Sensing the rhythmic behavior of nature, as expressed in its seasonal patterns, they felt that human lives and worldly events revolved around great wheels as well. Reincarnation of the soul and renewal of the world constituted twin expressions of the same fundamental principle. As Pythagoras, Plato, and other classical philosophers maintained, truth is everlasting and death an illusion. They considered the perfection of simple geometric shapes, particularly the circle, representative of the completeness and elegance of the universe.

Drawing from the notion that the cycles of stars and planets influenced the rhythms of human events, the pseudoscience of astrology emerged as one of the first forms of prophecy. Other types of supposed foreknowledge, similarly based on trying to decipher natural signs, included divination, the reading of auguries, the interpretation of dreams, and (above all in ancient Greece) the rendering of messages from oracles.

Eventually, starting in the Middle East and then spreading to many other parts of the world, the concept of linear time was born. In contrast to Greek and Eastern cycles, Jewish, Christian, and Islamic traditions embraced the idea that creation had a fixed beginning in time and that it will someday draw to a close. In prediction, mainstream religions struck a cautious tone, admonishing that the future was a slowly unraveling scroll known completely only by God. The unknowability, but inevitability, of death (and final judgment) became a vivid point of fear.

The story of the Fall, such an important part of all three cultures, served as a metaphor for the transition between the comfort of eternal repetition (the Garden of Eden), and the frightening responsibility of unidirectional change (the world outside the Garden). As religious scholar Mircea Eliade has described, when humankind began to accept that many aspects of the world were irreversible it left the innocence of natural perfection behind and set out into harsher terrain. Its only solace was the idea that by treading a single straight and narrow path, one might eventually regain paradise.

Starting in the twentieth century and continuing until today, humankind has experienced a second Fall, this time leaping from the narrow beam of simple linearity onto the tangled net of utter complexity. In the worlds of science and culture, unidirectional concepts have given way to labyrinthine, multiforked patterns. Absolute time and space have become replaced with the mosaic of relativity. Precise knowledge of physical parameters have become superceded by the quantum idea of simultaneous alternative histories. Fractal figures, such as strange attractors, have become new ways of describing dynamical systems—supplementing simple geometric patterns.

The computer revolution has accelerated these trends and extended them to methods of research and learning. Already, hypertext seems to be replacing (linear) text as the favored means of conveying detailed information. Like Theseus finding his way through the Minoan maze, researchers now navigate through complex webs of links seeking their hoped-for results. The new age of information has not only enabled the world to shrink, it has also allowed it to connect itself in ever more intricate patterns.

To process and interpret this wealth of data, the predictive sciences have struggled to keep pace. The requirements of the information age have tested the limits of conventional forecasting and opened up eager markets for new approaches. As Doyne Farmer reports, "Because we have so much information and the amount of information we can quantify is growing so rapidly, prediction is becoming more and more important—from people's preferences to climate change, to changes in the economy, to just about any kind of trend." [1]

Fortunately, in recent decades, scientific progress in the field of complexity has generated predictive models of a novel sort. For many types of endeavors, parallel processing methods have supplanted linear analyses as the most effective means of anticipating the future. These novel techniques, modeled on nature's evolutionary ability to shape itself to fit new circumstances, have offered a boon for forecasters in innumerable branches of the natural and social sciences.

The Prediction Revolution: Triumphs and Challenges

Twenty years ago, few specialists outside of the physical and mathematical sciences had even heard of chaos and complexity. Today, fractal dimensionality, information entropy, and other measures of chaotic behavior have entered the lexicons of physicians as well as physicists, mental health workers as well as mathematicians. State-space methods, involving plotting data in a special way and analyzing its complexity, have permitted diagnoses and treatment based on otherwise unseen patterns.

Consider, for example, the issue of predicting the onset of epileptic seizures. Affecting millions of people around the world, these unwanted intruders can lead to injurious falls, crashes and other accidents stemming from loss of bodily control. Until recently, no physician could reliably estimate the chances whether a seizure would soon occur. Experts searched in vain for characteristic signals, hoping to allay risks with drugs or electrical stimulation. Old-style linear thinking, however, proved inadequate for the task.

Neuroscientists J. Chris Sackellares and Leonidas Iasemedis of the University of Florida's Brain Institute decided to see if chaos theory would do the trick. Throughout the 1990s, they analyzed the EEG readings of patient after patient, attempting to detect preseizure states by means of mathematical analyses. They hoped that chaotic measures, such as Lyaponov exponents (an index of chaoticity) would provide warning flags, allowing physicians to head off seizures well before they affected patients. Over time, the researchers found that they could distinguish preseizure activity from normal brain function. They learned how to anticipate seizures minutes to hours in advance.

The breakthrough made by Sackellares and Iasemedis offers enormous hope to those suffering from epilepsy. Soon, the neuroscientists expect, implantable devices will be available for providing continuous detection and treatment of the conditions leading up to seizures. Rather than fearing chaos, they have conquered it, bringing forth one of the greatest triumphs of modern medical prediction.

Chaos in medicine is but one example in which the complex sciences have been successfully applied to forecasting. Scientists of all varieties have learned to tap into complexity's bounty. Neural networks, genetic algorithms, simulated annealing, and other methods for machine learning and optimization have provided valuable tools for a rapidly growing prediction industry. New scientific enterprises, based in corporate centers as well as academic venues, have taken on novel challenges in forecasting, from anticipating the ups and downs of the stock market to understanding the origins and possibilities of life.

A prediction contest, sponsored in 1991 by the Santa Fe Institute, highlighted the great strides made in the field. Contestants, drawn from assorted scientific disciplines, were presented with six time series patterns generated by various means, and were asked to extrapolate their results into the future as best they could. These graphs included representations of physiological data from a patient with sleep apnea, an account of currency exchange rate fluctuations, laser output records, white dwarf emission charts, the set of notes from a Bach fugue, and a specially constructed mathematical sequence. Such radically different patterns were chosen to minimize the chances of preconceptions. Some participants used newer predictive approaches, such as state-space reconstruction and neural networks, and others employed time-tested statistical techniques such as autoregression. In a triumph for more recent methods, based on complex systems and connectionist techniques, contestants who used them performed extremely well, producing far more accurate results than those using standard forms of extrapolation.[2]

Despite considerable progress made in all aspects of forecasting, in the past century, science has become increasingly and painfully aware of the fundamental limits of predictability. Foremost among these newly realized limitations is the blurring of information presented by the Heisenberg uncertainty principle. In the quantum world, no longer can all dynamic variables be said to be known for all times with exact precision. Gödel's theorem, and related theorems by Turing and Chaitlin, have set out similar bounds for mathematics. In arithmetic and geometry, as well as in atomic physics, incompleteness of knowledge is the rule rather than the exception. Science has come to realize that "there are always going to be inherently unpredictable aspects of the future."[3]

What can we know—or reasonably surmise—in advance? The answer to this question has challenged scientists and philosophers for millennia. We have made great strides in mapping the possibilities and limits of forecasting, but many dilemmas remain unresolved. Some of these pertain to mathematics. For example, how might one best understand the murky area between absolute certainty and pure chance? What is the true meaning of randomness? What fraction of mathematical problems simply have no answer?

Other questions involve the interplay between the mental and physical realms—an issue raised by Descartes but never (to my mind) satisfactorily resolved . Given that our images of the future depend on detailed mental models, one might wonder which features of the mind might be wholly replicated by machines? Which of the remaining properties are unique to humans? What are the mechanisms of free will (if one might describe them as mechanisms)? How does consciousness influence scientific results?

The relationship between humans and the cosmos raises yet another series of issues. How central or peripheral is intelligent life's place in the universe? Might we assume, as the Cosmological principle states, that all locales in space are essentially similar—and then use this conjecture to try to predict the future? How much of the cosmos—as the Anthropic principle ponders—would be different if humankind weren't here to observe it? And, the ultimate mystery: what lies beyond death—not just of ourselves, but of the universe as a whole?

Between Certainty and Chance

Predictability is a bridge resting on two solid pillars. One end of the span is supported by simple, regular mathematical models. Kepler's laws, Hooke's law (the principle that governs springs), the inverse-squared laws of electricity and gravitation, and other classical physical principles fall into this category, when they are applied to systems with few elements. For example, the behavior of the Moon, subject to the pull of Earth, can be forecast with great precision as a classic "two-body" gravitational problem.

The other end of the bridge rests on the statistical properties of pure randomness. Though it might seem strange to associate random numbers with prediction, many important scientific theorems are based on chance behavior. For instance, temperature and entropy, two of the fundamental variables in thermodynamics, are defined in terms of mathematical averages of the random motions of large groups of molecules. Typically, the more numerous a set of objects, the easier it is to render statistical judgments. Flip a fair coin one million times, and chances are, it would read tails in approximately 50 percent of its landings.

In between these two supports, however, the bridge of predictability can be quite shaky. The regions farthest from the extremes of simple, mechanistic systems on the one hand, and large, random ensembles on the other, are the most difficult to describe. Traditionally, scientists have tried to approximate these complex, nonrandom phenomena, as either small perturbations of the completely simple or minor variations of the wholly random. In some cases, such approximations work very well. Predicting the behavior of a spaceship traveling from Earth to the Moon—a "three-body" problem—is possible because the spaceship's mass might be considered negligible compared to those of Earth and the Moon.

In many other situations, though—the weather representing a prime example—assumptions of near-simplicity or near-randomness cannot be used. The equations that represent the weather do not constitute just one or two elements, nor do they depend on myriad randomly fluctuating variables. Rather, they are based on a small set of interacting quantities: air pressure, humidity, temperature, wind velocity, and so on. As Lorenz demonstrated, the resulting dynamics of such an intermediate model is neither periodic nor random, but rather the strange hybrid of deterministic chaos. Such behavior, though completely mechanistic in its governing equations, masquerades as randomness in its displayed activity, as depicted in the curiously twisted strange attractors found by Lorenz, Hénon, and others.

To scope out the murky domain between exactitude and chance, where perturbation theory does not apply, researchers have sought more flexible mathematical tools. Techniques for analyzing strange attractors and other aspects of chaos represent an important start for such efforts. Neural networks, with their ability to master the nuances of novel systems, constitute another critical stride forward. Far more of this terra incognita calls for exploration—offering adventurous prospects for twenty-first-century mathematicians and scientists.

Minds, Machines, and Mathematics

The brain's systems for learning and prediction are unsurpassed in their flexibility. Using what others say or what the senses reveal in constructing detailed internal representations of the world, manipulating these mental models and thereby predicting what might happen next, comprise abilities of phenomenal sophistication. And these are only some of the talents that each of us carry in our heads.

True, our predictive skills are limited by what our senses take in and by our ability to process this information quickly and effectively. We cannot always remember what we see or hear, let alone understand it and use it to anticipate the future. As psychologist Philip Johnson-Laird has pointed out, our representations are necessarily simplified versions of the real world. Their shadow plays sometimes belie more solid truths.

Computers have the capacity to record and process enormous streams of facts. Consequently, in certain areas of forecasting involving algorithmic "crunching" of known data, they perform exceedingly well. Give a machine a formula, or the information required to establish a formula, and it can spew out future values indefinitely.

Many predictive problems require insight, however, rather than simple churning out of facts. In this domain, humans excel and machines often fail. Insight typically entails creating an internal representation of an issue and using it as a springboard to make a mental leap. Who hasn't thought about a particular dilemma before going to sleep, and then woken up knowing the solution? Computers, in contrast, do what they are programmed to do. They could arrive at an answer only if their programmers offered them means of reaching such a conclusion—either directly or through a learning algorithm.

Some artificial intelligence (AI) experts assert that computers will someday possess intuitive abilities at least equal to humans. They argue that in certain limited realms, such as chess, computers can already mimic human insight by examining millions of possibilities and then making judgments accordingly. Eventually, these AI proponents contend, computers will be so proficient in all such aspects of decision making that nobody would be able to distinguish their behavior from humans.

However, machines operate by means of step-by-step processes. As Turing, Chaitlin, and others have pointed out, certain mathematical problems cannot be resolved through algorithmic methods in a finite amount of time. Resolving such computationally irreducible questions would elude even the fastest, most powerful computers.

Human brains, on the other hand, can skip steps if necessary. They can avoid "infinite loops" by switching gears; if a problem seems intractable, maybe another approach would work out better. Moreover, they can envision issues holistically as well as in component form. In short, they are instruments capable of solving problems in many different ways, not just through algorithms—freely altering strategies whenever necessary. They exhibit free will, a feature a preprogrammed instrument cannot muster by definition.

The Enigma of Free Will

Why is it often so much harder to predict human events than to anticipate natural occurrences? Unlike the Moon revolving steadily in its path around Earth, or a ball dropping unequivocally off a steep tower, a person can change his mind in midcourse and, if physically possible, alter his actions as well. Even the most hardened criminal might someday repent. Even the most kind-hearted individual might render a deadly decision. Those who think that by knowing the boy they can predict the man often fool themselves; truly no one anticipated the genius of Einstein or the evil of Hitler.

Free will comprises one of life's great mysteries. It is neither wholly random, nor completely predetermined. If you believe that it is either, then quickly think of a number from one to ten. Did you employ a random number generator or spin a wheel to produce your answer? Of course not. Did you follow a deterministic algorithm, one that anyone else in the world could decipher? No again. Did you *have to* think of a number; could you have stood on your head and recited limericks instead? Naturally, you could have thought and done anything (within your mental and physical capacity). Might someone beyond the human realm (an omniscient being, perhaps) understand the laws of the physics and the workings of the brain well enough to predict everything? Conceivably yes. However, as Popper has shown, no one on Earth might do so. Even if one could begin to decipher another's complete set of thoughts and experiences—a doubtful task indeed—it would take so long to do so that the subject of such an experiment could make countless other decisions in the interim.

Some scientists—Bohr most famously—have asserted that free will is a quantum phenomenon. Human choice, they assert, is akin to the decision an electron makes when it collapses into the wavefunction representing a certain spin. But it seems to me that although the average physical properties (expectation values) of subatomic particles fall into precise patterns—the energy levels of hydrogen, for instance—conscious intelligence displays no such standard, overall features.

Other theorists have characterized free will as an emergent phenomenon, stemming from the apparent randomness of deterministic systems. Just as no one might forecast the weather one year in advance, no one might anticipate someone else's (or even their own) thoughts one year (or indeed any time) ahead. True, chaotic behavior and human actions are each essentially unpredictable. Yet until someone shows that the mental activity, on a fundamental level, obeys certain objective "mechanistic" principles—and no one, so far, has come close—equating human thought with the actions of automata seems to me to be making a titan's leap.

Then what hope do we have of gauging the shape of things to come? Much in nature, but unfortunately far less in human life. Free will's open agenda even makes it hard to conceive how the most speculative of prognosticative ideas—the notion of traveling to different times and then returning—might ever be accomplished.

Time travel is a long-held dream, featured in thousands of fictional works. It would be wonderful to venture into past eras or future epochs, and then come back to report their otherwise unreachable marvels. Yet round-trip journeys through time would likely lead to untold paradoxes. Because we are not preprogrammed automata, if we went to other eras, our ability to freely alter our environments would disrupt destiny's steady stream, resulting in contradictions when we returned. Even backward communication, via "tachyonic antitelephones" would disrupt reality's fabric. Only strictly future-oriented time travel, in which the decisions rendered by free will would follow the natural direction of cause and effect, would be immune to the ailments of paradox.

One of the curious things about free will, however, is that although it is open, it is guided by experiences and motivations. If asked to pick a number from one to ten, and then offered one million dollars for picking "three," most people would make that choice. Hence, in certain domains, by understanding what drives decisions, one might venture guesses as to what might happen. Our common perceptions as human beings enable us to make such judgments. Though we cannot enter others' minds, we might imagine what they are feeling. Even if these hunches are correct only some of the time, if one guesses even slightly better than one's competitors, one might gain an advantage. Hence the great market for stock market forecasting—which essentially uses models as to why and when people buy and sell to gain enough advantage to make a profit. Superior intuitions about what others might be thinking also aid in social settings; those "tuned-in" to their companions' feelings naturally find better response. When such instincts are absent, as in the case of severe autism, successful social interactions become nearly impossible.

In summary, free will means that prediction in the human realm can never be perfect. Because of such uncertainty, long-term social forecasts generally miss the mark. If one develops superior mental models of what motivates human decision making, however, one might, in certain cases, gain a predictive advantage over competitors—critical in everything from playing the stock market to finding friendship and romance.

Our Place in the Cosmos

On Earth, free will, self-awareness, and other aspects of higher intelligence uniquely define our species. No other terrestrial life forms possess these gifts. In the earthly realm, humankind sits upon the highest throne, steering the course of planetary affairs.

With regard to the cosmos in general, we cannot, however, make such claims. There may well be other species like us. Although current searches for extraterrestrial intelligence have come up empty, recent discoveries of planets around other stars suggest that worlds like ours may be common in space. Considering the potentially vast number of planets in the universe, it is likely that at least some of them harbor other cognizant beings.

Belief that humankind might be typical is a fairly recent phenomenon. For much of history, humankind thought it occupied a special place in the cosmos. When Copernicus redesigned astronomical maps, removing Earth from its central location, he initiated a grand rethinking of humanity's place in creation. Over the centuries that followed, numerous discoveries showing how vast the universe truly is and how tiny our world is by comparison fortified the premise that we are but beings on an average world in an average position in space.

The Cosmological principle, a statement of our averageness, has become an important aspect of astrophysical prediction. Cosmologists often use it to infer that if something is true for Earth, it must be true, on average, for the cosmos in general. For instance, because the universe looks isotropic to us, they surmise that it appears the same everywhere. The apparent homogeneity of space justifies, in turn, the Friedmann models of cosmic origin and destiny.

The more controversial Anthropic principle has also found use in forecasting. Some researchers argue that our presence as sentient observers of the universe helps pinpoint its properties. They then employ these narrowed-down specifications to justify models of its past and future.

Cosmologist J. Richard Gott has fused these ideas into a statement that he calls the Copernican principle. Combining notions of averageness and uniqueness, he states that "the location of your birth in space and time in the universe is privileged (or special) only to the extent implied by the fact you are an intelligent observer." Furthermore, unless shown otherwise, you should assume that "your location among intelligent observers is not special but rather picked at random."[4]

Gott has used his principle for all manner of predictions, from estimating the odds of future political and social events to calculating when the world will likely end. He bases his guesses on the assumption that one generally occupies an average place in time, unless otherwise demonstrated. Consequently, for any event of finite duration (Earth's existence, for instance), one is most likely to be living during the middle of its time span, and least likely to be living during the beginning or end. From this assumption of averageness and knowledge of how long something has already existed, Gott estimates how much longer it will last. This provides him with extremely broad, but often accurate guesses.

In 1993, he applied the Copernican principle to predict how long certain Broadway plays would run. He looked up all the plays that were running in New York theaters, and forecast when each would close. For example, he predicted that *Cats,* which had then been on Broadway for 3,885 days, would last between 100 days and 414 years. According to him, as of summer 1999, he was "batting a thousand"; that is, every play he thought would end its run, did indeed close on schedule.[5]

Admittedly, Gott's method is far from precise, but it cleverly uses intuition about the world to anticipate its future properties. No machine, programmed to analyze events step by step, could have developed such a strategy unless it had already been so instructed.

Intuition can carry one only so far. There are certain domains that elude attempts at understanding, such as the mysteries of life and death, universal as well as personal. What, if anything, will follow the end of time as we know it? Does the cosmos have an ultimate purpose, and will we ever fathom it? These religious and philosophical questions, considered by many outside the scope of science, constitute the final frontiers of prediction.

The Essential Mystery

As the writer Borges has pointed out, the ultimate labyrinth is a wholly featureless plain. Nothing is more featureless than the solitary wastelands of death. Beyond the realm of the familiar and the domain of the predictable lies a mysterious kingdom, inaccessible to human comprehension. Fate conspires to block us from peering into these hidden lands until it is too late to report back what we have found.

Through our senses, we are privy to much information about the world. By use of our natural abilities to fill in what is unseen and unfelt, we can often construct mental representations of what eludes direct perception. Our capability, for instance, to construct moving images from a series of still shots serve as testimony to the reconstructive powers of our minds.

Yet our cognitive abilities can carry us only so far. Some concepts lie simply beyond human grasp. Without proper footholds to help us hoist ourselves above the rocky walls of our immediate experiences, prediction, in such areas, becomes impossible. To unravel the secrets of life and death, to fathom free will and consciousness, to understand temporal passage requires, perhaps, greater understanding than our limited minds can muster.

What lies on the other side of time's great divides—the beginning and demise of the cosmos as we know it to be? Are there higher dimensions and parallel worlds, and if so, what strange entities might inhabit them? Why are we drawn inevitably toward the future, forever leaving the past behind? If each effect has a cause, what was the first? These are questions for which destiny's riddle has no answer.

Fortunately, there is much about the world that we can understand, and much of the future that we can anticipate. Science provides us with the tools to model diverse types of natural phenomena and to project our results far into times to come. With the advent of computers and the development of enhanced techniques for learning, optimization, and forecasting, our ability to predict the future is the greatest it has ever been. Though today we face predictive limits posed by machines, methodology, mathematics, and mortality itself, the future holds great promise for us all.

FURTHER READING

Adams, Fred, and Greg Laughlin. 1999. The Five Ages of the Universe: Inside the Physics of Eternity. New York: Free Press.

Asimov, Isaac. 1966. *Foundation*. New York: Avon Books.

Barrow, John D, and Frank J. Tipler. 1988. *The Anthropic Cosmological Principle*. New York: Oxford University Press.

Bass, Thomas A. 1986. *The Eudaemonic Pie*. New York: Vintage Books.

Bellamy, Edward. 1996. *Looking Backward*. New York: Dover Publications.

Borges, Jorge Luis. 1967. "The Aleph." In *A Personal Anthology,* ed. and trans. Anthony Kerrigan. New York: Grove Weidenfeld.

Box, George E. P., and Gwilym M. Jenkins. 1976. *Time Series Analysis: Forecasting and Control.* San Francisco: Holden-Day.

Capek, M. 1961. *The Philosophical Impact of Contemporary Physics*. Princeton, N.J.: D. Van Nostrand.

Casti, John L. 1990. *Searching for Certainty : What Scientists Can Know about the Future*. New York: W. Morrow.

Casti, John L., and Anders Karlqvist, eds. 1991. *Beyond Belief: Randomness, Prediction and Explanation in Science*. Boston, CRC Press.

Cicero, Marcus Tullius. 1964. *De Divinatione*, trans. William Falconer. Cambridge, Mass.: Harvard University Press.

Crease, Robert P., and Charles C. Mann. 1986. *The Second Creation: Makers of the Revolution in 20th-Century Physics*. New York: Collier.

Dalkey, Norman. 1969. *The Delphi Method*. Santa Monica, Calif.: Rand Corp.

Deutsch, David. 1997. *The Fabric of Reality*. New York: Penguin.

Drosnin, Michael. 1997. *The Bible Code*. New York: Simon and Schuster, 1997.

Eco, Umberto. 1988. *Foucault's Pendulum,* trans. William Weaver. New York: Harcourt Brace Jovanovich.

Ferris, Timothy. 1999. "How to Predict Everything." In *The New Yorker*, July 12, 33–37.

Feynman, Richard P. 1985. *Surely You're Joking Mr. Feynman*. New York: Bantam Books.

Gardner, Martin. 1983. *Wheels, Life, and Other Mathematical Amusements.* New York: W. H. Freeman.

Gardner, Martin. 1985. *The Whys of a Philosophical Scrivener.* New York: Oxford University Press.

Gardner, Martin. 1986. *Knotted Doughnuts and Other Mathematical Entertainments.* New York: W. H. Freeman.

Gardner, Martin. 1986. *The Unexpected Hanging and Other Mathematical Diversions.* New York: Simon and Schuster.

Gardner, Martin. 1988. *Time Travel and Other Mathematical Bewilderments.* New York: W. H. Freeman.

Gleick, James. 1987. *Chaos: Making a New Science.* New York: Viking.

Goldberg, David E. 1989. *Genetic Algorithms in Search, Optimization, and Machine Learning.* Reading, Mass.: Addison-Wesley.

Halpern, Paul. 1990. *Time Journeys: A Search for Cosmic Destiny and Meaning.* New York: McGraw-Hill.

Halpern, Paul. 1992. *Cosmic Wormholes: The Search for Interstellar Shortcuts.* New York: Dutton.

Halpern, Paul. 1998. *Countdown to Apocalypse: Asteroids, Tidal Waves and the End of the World.* New York: Plenum.

Hamilton, Edith. 1993. *The Greek Way.* New York : W. W. Norton and Co.

Hebb, Donald O. 1949. *The Organization of Behavior: A Neuropsychological Theory.* New York: Wiley.

Helmer, Olaf. 1983. *Looking Forward: A Guide to Futures Research.* Beverly Hills, Calif.: Sage Publications.

Holland, John H. 1998. *Emergence: From Chaos to Order.* Reading, Mass.: Helix.

Huxley, Aldous. 1965. *Brave New World and Brave New World Revisited.* New York: Harper and Row.

Johnson-Laird, Philip N. 1983. *Mental Models.* Cambridge, Mass.: Harvard University Press.

Kahn, Herman, and Anthony Wiener. 1967. *The Year 2000: A Framework for Speculation on the Next Thirty-Three Years.* New York: Macmillan.

Kauffman, Stuart A. 1993. *The Origins of Order: Self-Organization and Selection in Evolution.* New York: Oxford University Press.

Keynes, John Maynard. 1972. "Newton, the Man." *The Collected Writings of John Maynard Keynes: Volume X, Essays in Biography* . Cambridge: Macmillan, 366–370.

Koestler, Arthur. 1959. *The Sleepwalkers: A History of Man's Changing Vision of the Universe.* New York: Macmillan.

Leslie, John. 1996. *The End of the World.* London: Routledge.

Levy, Stephen. 1992. *Artificial Life: A Report from the Frontier Where Computers Meet Biology.* New York: Vintage Books.

Pagels, Heinz. 1988. *The Dreams of Reason : The Computer and the Rise of the Sciences of Complexity.* New York: Simon and Schuster.

Plato. 1949. *Timaeus,* trans. Benjamin Jowett. Indianapolis: Bobbs-Merrill.

Popcorn, Faith. 1991. *The Popcorn Report: Faith Popcorn on the Future of Your Company, Your World, Your Life.* New York: Doubleday.

Popper, Karl R. 1982. *The Open Universe: An Argument for Indeterminism.* Totowa, N.J.: Rowman and Littlefield.

Randi, James. 1993. *The Mask of Nostradamus: The Prophecies of the World's Most Famous Seer.* Buffalo, N.Y.: Prometheus Books.

Rescher, Nicholas. 1998. *Predicting the Future : An Introduction to the Theory of Forecasting.* Albany, N.Y.: State University of New York Press.

Sagan, Carl. 1980. *Cosmos.* New York: Ballantine Books.

Scholem, Gershom. 1974. *Kabbalah.* New York: Quadrangle/New York Times Book Co.

Smolin, Lee. 1997. *The Life of the Cosmos.* New York: Oxford University Press.

Toffler, Alvin. 1970. *Future Shock.* New York: Random House.

Waldrop, M. Mitchell. 1992. *Complexity: The Emerging Science at the Edge of Order and Chaos.* New York: Touchstone Books.

Weigend, Andreas, and Neil Gershenfeld, eds. 1992. *Time Series Prediction: Forecasting the Future and Understanding the Past.* Reading, Mass.: Addison-Wesley.

Wells, H. G. 1960. "The Time Machine." In *Three Prophetic Novels,* selected by E. F. Bleiber, New York: Dover.

NOTES

Introduction

1. Martin Gardner, personal communication, January 1, 1999.

Chapter 1

1. John Hale in Bob Holmes, "Did Ancient Gases Inspire Powers of Prophecy?" *New Scientist* (3 May 1997): 16.

2. Plato, *Timaeus,* trans. Benjamin Jowett (Indianapolis: Bobbs-Merrill, 1949).

3. Arthur Koestler, *The Sleepwalkers: A History of Man's Changing Vision of the Universe* (New York: Macmillan, 1959), 56.

4. Reported in Marcus Chown, "In the Shadow of the Moon," *New Scientist* 161, 2171 (30 January 1999) 30.

5. Marcus Tullius Cicero, *De Divinatione* 2, ix, 23, trans. William Falconer, Cambridge, Mass.: Harvard University Press, 1964.

6. Cicero, *De Divinatione,* 399.

7. Cicero, *De Divinatione,* 465.

8. See Paul Halpern, *Countdown to Apocalypse* (Cambridge, Mass.: Perseus, 1998), for a more detailed discussion of apocalyptic movements and beliefs.

9. James Randi, phone interview, 29 December 1998.

Chapter 2

1. Max Caspar, *Johannes Kepler* (Stuttgart: Kohlhammer, 1948), 117. Quoted in Arthur Koestler, *The Sleepwalkers* (New York: Macmillan, 1959), 304.

2. Tycho Brahe, quoted in Carl Sagan, *Cosmos* (New York: Ballantine Books, 1980), 45.

3. Johannes Kepler, "Preamble to the Table of Contents," *Astronomia Nova* 3. Quoted in Arthur Koestler, *The Sleepwalkers* (New York: Macmillan, 1959), 315.

4. Galileo Galilei, "Letter to the Grand Duchess Christina." Quoted in Arthur Koestler, *The Sleepwalkers* (New York: Macmillan, 1959), 437.

5. John Maynard Keynes, "Newton, the Man," in *The Collected Writings of John Maynard Keynes: Volume X, Essays in Biography* (Cambridge: Macmillan, 1972), 366.

6. Pierre Simon Laplace, quoted in M. Capek, *The Philosophical Impact of Contemporary Physics* (Princeton, N.J.: D. Van Nostrand, 1961), 122.

7. James Gleick, *Chaos: Making a New Science* (New York: Viking, 1987), 321.

8. John Dewey and William James, quoted in Martin Gardner. "Newcomb's Paradox," *Knotted Doughnuts and Other Mathematical Entertainments* (New York: W. H. Freeman, 1986), 155–156.

9. Charles Brenner, *An Elementary Textbook of Psychoanalysis* (New York: International Universities Press, 1955), 12.

Chapter 3

1. H. G. Wells, "The Time Machine," in *Three Prophetic Novels,* selected by E. F. Bleiber (New York: Dover, 1960), 5.

2. Jorge Luis Borges, "The Aleph," in *A Personal Anthology,* ed. and trans. Anthony Kerrigan (New York: Grove Weidenfeld, 1967), 147.

3. Martin Gardner, "The Paradox of the Unexpected Hanging," in *The Unexpected Hanging and Other Mathematical Diversions* (New York: Simon and Schuster, 1986), 12.

4. Martin Gardner, "Newcomb's Paradox," in *Knotted Doughnuts and Other Mathematical Entertainments* (New York: W. H. Freeman, 1983), 157.

5. Stephen W. Hawking, "Chronology Protection Conjecture,"*Physical Review* D46 (15 July 1992): 603.

6. Ibid., 610.

7. J. Richard Gott III and Li-Xin Li, "Can the Universe Create Itself?" *Physical Review* D 58 (1998): 23501.

Chapter 4

1. David Deutsch, *The Fabric of Reality* (New York: Penguin Books, 1997), 263.

2. Richard P. Feynman, *Surely You're Joking Mr. Feynman* (New York: Bantam Books, 1985), 48.

3. Interview with Richard Feynman, 1985, in Robert P. Crease and Charles C. Mann, *The Second Creation: Makers of the Revolution in 20th-Century Physics* (New York: Collier Books, 1986), 138.

Chapter 5

1. J. Doyne Farmer, quoted in Thomas A. Bass, *The Eudaemonic Pie* (New York: Vintage Books, 1986), 35.

2. J. Doyne Farmer, phone interview, 15 June 1999.

3. J. Doyne Farmer, quoted in Thomas A. Bass, *The Eudaemonic Pie* (New York: Vintage Books, 1986), 191.

4. John L. Casti, "Chaos, Gödel and Truth," in *Beyond Belief: Randomness, Prediction and Explanation in Science*, ed. John L. Casti and Anders Karlqvist. (Boston, CRC Press, 1991), 290.

5. J. Doyne Farmer, phone interview, 15 June 1999.

6. Gregory Chaitlin, quoted in John L. Casti, "Chaos, Gödel and Truth," in *Beyond Belief: Randomness, Prediction and Explanation in Science,* ed. John L. Casti and Anders Karlqvist (Boston, CRC Press, 1991), 316.

Chapter 6

1. John Conway, quoted in Steven Levy, *Artificial Life: A Report from the Frontier Where Computers Meet Biology* (New York: Vintage Books, 1992), 52.

2. John Holland, personal communication, 28 May 1999.

3. John Holland, reported in M. Mitchell Waldrop, *Complexity: The Emerging Science at the Edge of Order and Chaos* (New York: Touchstone Books, 1992), 145–146.

4. Philip N. Johnson-Laird, *Mental Models* (Cambridge, Mass.: Harvard University Press, 1983), 10.

5. James Gleick, *Chaos: Making a New Science* (New York: Viking, 1987), 283.

6. Miodrag Dodic, reporting in Karen Schmidt, "Programmed at Birth," *New Scientist* 163, 2195 (17 July 1999): 27.

Chapter 7

1. J. Doyne Farmer, phone interview, 15 June 1999.

2. J. Doyne Farmer, "Market Force, Ecology and Evolution," unpublished manuscript, 21.

3. Nicholas Rescher, *Predicting the Future: An Introduction to the Theory of Forecasting* (New York: State University of New York Press, 1998), 12.

4. J. Doyne Farmer, phone interview, 15 June 1999.

5. J. Doyne Farmer, phone interview, 15 June 1999.

6. Isaac Asimov, *Foundation* (New York: Avon Books, 1966), 17.

7. From the World Future Society membership guide.

8. Herman Kahn and Anthony Wiener, *The Year 2000: A Framework for Speculation on the Next Thirty-Three Years* (New York: Macmillan, 1967), 234.

9. Ibid.

10. Ibid., 245.

11. Ibid., 246.

12. Faith Popcorn, *The Popcorn Report: Faith Popcorn on the Future of Your Company, Your World, Your Life* (New York: Doubleday, 1991), 187–188.

13. Ibid., 131.

Chapter 8

1. Saul Perlmutter, personal communication, 15 January 1998.

Conclusion

1. J. Doyne Farmer, phone interview, 15 June 1999.

2. Andreas Weigend and Neil Gershenfeld, "The Future of Time Series: Learning and Understanding," in Andreas Weigend and Neil Gershenfeld, eds., *Time Series Prediction: Forecasting the Future and Understanding the Past* (Reading, Mass.: Addison-Wesley, 1992), 7.

3. J. Doyne Farmer, phone interview, 15 June 1999.

4. J. Richard Gott, "Implications of the Copernican principle for our future prospects," *Nature* (27 May 1993): 316. Quoted in John Leslie, *The End of the World* (London: Routledge, 1996), 16.

5. J. Richard Gott, reported in Timothy Ferris, "How to Predict Everything," *The New Yorker* (12 July 1999): 35.

INDEX